EXTREME MINING MACHINES
STRIPPING SHOVELS AND WALKING DRAGLINES

KEITH HADDOCK

MBI Publishing Company

First published in 2001 by MBI Publishing Company, 729 Prospect Avenue, PO Box 1, Osceola, WI 54020-0001 USA

© Keith Haddock, 2001

All rights reserved. With the exception of quoting brief passages for the purposes of review, no part of this publication may be reproduced without prior written permission from the Publisher.

The information in this book is true and complete to the best of our knowledge. All recommendations are made without any guarantee on the part of the author or Publisher, who also disclaim any liability incurred in connection with the use of this data or specific details.

We recognize that some words, model names and designations, for example, mentioned herein are the property of the trademark holder. We use them for identification purposes only. This is not an official publication.

MBI Publishing Company books are also available at discounts in bulk quantity for industrial or sales-promotional use. For details write to Special Sales Manager at Motorbooks International Wholesalers & Distributors, 729 Prospect Avenue, PO Box 1, Osceola, WI 54020-0001 USA.

Library of Congress Cataloging-in-Publication Data Available
ISBN 0-7603-0918-3

On the front cover: This Marion 5761 stripping shovel used a 65-cubic yard dipper on a 170-foot boom. The operating weight of this monster was 7,575,000 pounds. The shovel was fitted with eight crawler assemblies, measuring 23 feet long, 5 feet wide and weighing 85,000 pounds each.
Eric C. Orlemann collection

On the frontispiece: The original bucket of the largest stripping shovel ever made, the Marion 6360, takes a 180 cubic yard bite out of the earth. The dipper was equipped with two doors, each weighing 15 tons.
Eric C. Orlemann collection

On the title page: This rare photo captures the ultimate walking dragline—*Big Muskie* in action. *Muskie*, Bucyrus Erie model 4250-W, swings the biggest bucket (220 cubic yards) ever to be suspended on a boom. *Eric C. Orlemann*

On the back cover, top: The 8800 series was the world's largest dragline when ordered from Marion in 1961 for Peabody Coal Company. It carried a 85 cubic yard bucket on a 275 foot boom, and it weighed an astounding 6,285 tons. *Eric C. Orlemann* *Bottom:* Two record-beaters in the same cut! The Marion 6360 "Captain" shovel is digging the lower overburden and the parting between two seams of coal while the massive 5872-WX cross-pit bucket wheel excavator, largest of its type, removes the upper overburden. The machines are digging from right to left. *Eric C. Orlemann*

ACKNOWLEDGMENTS

I would like to give special thanks to the many individuals and organizations who have helped to make this publication not just another picture book, but one filled with accurate data and spectacular images of the colossal machines of the earthmoving world. Those listed below, and others too numerous to mention, have supported me in my endeavors. I really appreciate your time and efforts. A special thanks to my friend and fellow author Eric Orlemann for his guidance and for providing many of his superb images. Most of all, I thank my wife Barbara for her support and help in editing the text, and also her tolerance during the many hours and days when I was lost to the computer.

- Bill Williams, P&H Mining Equipment
- Bob Jelinek, Bucyrus International, Inc.
- Bruce Knight, Transwest Dynequip
- Dave Wootton, Hednesford, England
- David Lang, Bucyrus International, Inc.
- Denis Gaspe, Fording Coal Ltd.
- Dimitrie Toth Jr., Waterford, Michigan, U.S.A.
- Don Frantz, Historical Construction Equipment Association
- Eric Orlemann, ECO Communications
- Jim Rosso, Bucyrus International, Inc.
- Peter Grimshaw, PNG Communications, England
- Peter Gilewicz, Parker Bay Company
- Tim Backus, P&H Mining Equipment
- Mark Dietz, P&H Mining Equipment
- Gregg Yaunke, P&H Mining Equipment
- Heinz-Herbert Cohrs, Holstein, Germany
- Urs Peyer, Brunnen, Switzerland
- Tom Berry, Historical Construction Equipment Association
- Dale Davis, Cadiz, Ohio, U.S.A.
- Walter Bennett, Scranton, Pennsylvania, U.S.A.

Edited by Kris Palmer
Designed by Mark Odegard

Printed in China

Illustration on following page.
This is where it all started! The first mechanical excavator was the steam shovel invented by William S. Otis in 1835. These early shovels did not fully revolve. Only the front end swung through about 180 degrees. For many decades, they were used on railroad construction, and they became known as "railroad shovels."
—Keith Haddock collection

CONTENTS

PART ONE
STRIPPING SHOVELS

CHAPTER ONE
**Stripping Shovel
Origin & Application**
7

CHAPTER TWO
Stripping Shovel Chronology
17

CHAPTER THREE
The Super Strippers
35

**PART TWO
WALKING DRAGLINES**

CHAPTER FOUR
**Walking Dragline
Origin & Application**
57

CHAPTER FIVE
Walking Dragline Chronology
79

CHAPTER SIX
The Giant Draglines
109

APPENDIX 1
A Brief Manufacturers' History
121

APPENDIX 2
Named Machines
124

APPENDIX 3
Table of Stripping Shovels
125

APPENDIX 4
Table of Walking Draglines
126

Index
128

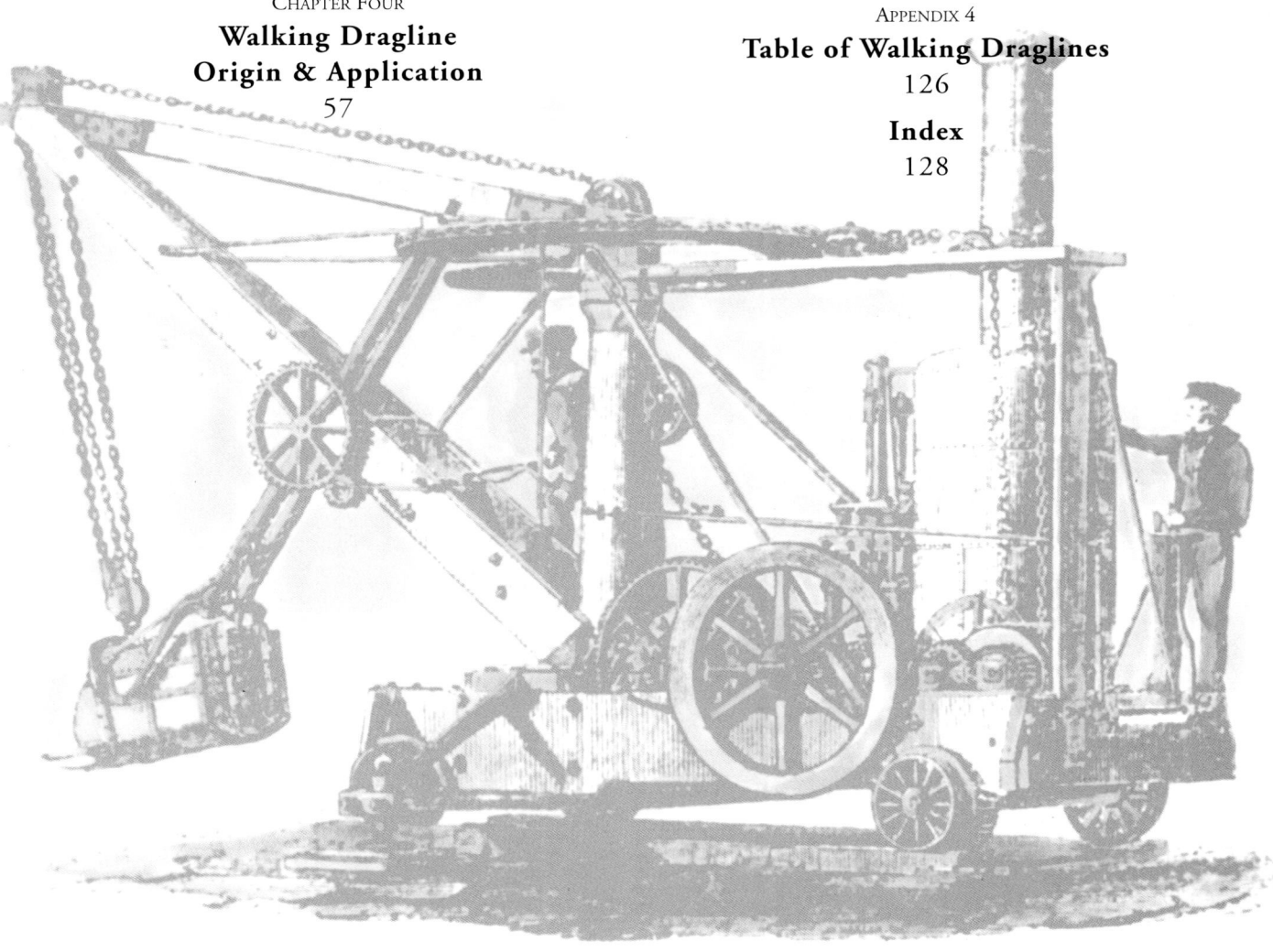

PART ONE
STRIPPING SHOVELS

CHAPTER ONE

STRIPPING SHOVEL ORIGIN & APPLICATION

Stripping shovels are the giants of the earthmoving industry. They belong to the family of machines known as excavators, and they move vast quantities of earth and rock—often 24 hours a day, seven days a week. Stripping shovels are king of the mobile machine world, and include the 15,000-ton *Captain* shovel, the heaviest piece of equipment ever to move under its own power on land (see chapter 3). Only the big ocean ships have a greater moveable mass than these monster machines.

A new Marion 5761 stripping shovel starts to uncover coal at the Ayrgem Mine, Kentucky, in 1970. Marion built 16 of the very successful 5761 shovels from 1959 to 1971. Bucket capacity ranged from 60 to 75 cubic yards, with an operating weight of 3,800 tons.
—Eric C. Orlemann collection

The first stripping shovels were mounted on rails. This electric example was built by Ruston Hornsby Ltd. in England. It's a No. 135, carrying a 4-cubic-yard dipper, at work in a brickworks clay pit.

—Keith Haddock

Stripping shovels work in surface mines or large quarries, not generally visible to the public. But if one happens to be working close to a public road, this mechanical monster is likely to attract a large group of onlookers. Stripping shovels strip away earth and rock (overburden) covering valuable minerals such as coal, ironstone, or limestone. They work in the same way as regular cable shovels, but unlike their smaller counterparts, they don't need to load another vehicle to carry away the material. Stripping shovels are necessarily built to huge proportions, giving them a much greater range. This allows them to dig from a high face, swing round, and dump the material well clear of the working area.

During operation the stripping shovel digs long strips or cuts, and casts the overburden into the neighboring cut from which the mineral has already been removed. It's like plowing a field where the furrow being dug is turned over to the adjacent furrow, but of course on a much larger scale. The stripping shovel is positioned on top of the mineral it is uncovering, and it moves forward as the digging face advances.

Because of the huge costs and the engineering and manufacturing challenges involved, only a very small number of companies have ventured into designing and building giant stripping shovels. In the United States, two companies have dominated manufacture: Marion Power Shovel Company and Bucyrus-Erie Company. The latter, now known as Bucyrus International, Inc., purchased Marion Power Shovel in 1997. The merger of these two companies ended a great rivalry that began with the founding of the Marion Steam Shovel Company in 1884. This rivalry was most intense in the area of stripping shovels. Both companies vied repeatedly for the latest innovation, or the largest machine (see chapter 2).

With a few exceptions, the Marion and Bucyrus companies have supplied the entire world market for large stripping shovels. Several other manufacturers did produce this type of machine in the early part of the twentieth century. They included Whitaker, Grossmith, and Taylor & Hubbard in England, and Browning in the United States. In addition, Ransomes & Rapier Ltd. built stripping shovels in England, starting in 1934 and ending in 1943. A history of Rapier stripping shovels appears in chapter 2, and their machines are listed in Appendix 3. The German firm Menck built one in the 1920s, and a few have been built in the former Soviet Union for use in that country. But there was never any serious competition for Marion and Bucyrus, whose names will forever be synonymous with stripping shovels.

A factory shot at Ruston works in Lincoln, England. A No. 135 steam-powered crawler excavator is under erection, destined for export to India.

—Keith Haddock

One of the few stripping shovel competitors to Marion and Bucyrus was the British firm, Ransomes & Rapier Ltd. This machine is an early rail-mounted Model 5360, uncovering ironstone in central England. The 650-ton machine carried an 8-cubic-yard dipper on a 104-foot-long boom.
—Keith Haddock collection

The big stripping shovels are handled with care by their operators. After all, the big stripper is often the main machine at the mine, and the mine's entire output depends on the shovel running properly during scheduled operation. If the shovel is down, the mine is not making money, and the workers' livelihood is at stake. So as far as budgets allow, maintenance is carried out on a regular basis to keep the machine in top working condition.

Like ships on the ocean, most of the large stripping shovels and walking draglines are given names by their owners (see Appendix 2). Very often the name is chosen from those submitted in a contest. The employees, or children at a school close to the mine, submit suggestions. The machines are regarded affectionately by those who work on and around them. Each machine has its own personality, and many a retiree will recall with pride stories and events involving the machine he worked on—sometimes for his entire working life.

MACHINE DESCRIPTION

The giant stripping shovels, like the smaller shovels, consist of an upper frame carrying the hoist, crowd, and swing machinery. This frame also carries the gantry and boom, and is mounted on a turntable on top of a lower frame. The early stripping shovels utilized the hoist machinery in the upper deck to power the propel machinery also. A jaw clutch could switch the power from one function to the other. The first machines were rail-mounted, but crawler tracks were added in the 1920s.

The largest stripping shovels are mounted on an eight-crawler lower frame, but some smaller examples were mounted on four crawlers. In addition, many of the smaller two-crawler cable-operated shovels could be fitted with longer-than-standard booms allowing them to perform stripping shovel duties in relatively shallow overburden. These types of stripping shovels were most popular in the 1940s and 1950s in the shallow overburden coal regions of Ohio, West Virginia, and western Pennsylvania.

Many excavator manufacturers who did not normally supply excavators to the surface mining industry offered the long-boom option. Thus, many makes of smaller stripping shovels, such as Osgood, Lorain, and Northwest, were seen working in the coal fields. Of course, the dipper capacity of these machines was less than could be carried by the same shovel fitted with its regular boom. The extra reach and the elimination of haulage vehicles offset the smaller dipper

The smallest Rapier stripping shovel was the 5160, weighing 258 tons, of which only two were built. The electric-powered 3-1/2-cubic-yard machine featured an unusual four-crawler undercarriage. The machine shown here is working in a limestone quarry in central England.
—Keith Haddock

capacity, however, and these machines proved profitable for their owners.

SHOVEL MOUNTINGS

The very first stripping shovels were actually steam railroad shovels fitted with a long boom. The earliest documented models were made by Vulcan in 1899. William S. Otis invented the very first steam shovel in 1835 for use on railroad construction, and the standard railroad shovel was developed from his design. Railroad shovels were most popular in the nineteenth century and the first decade of the twentieth century. They gained their name from their popular use on railroad construction, and because they were usually mounted on standard railroad tracks. Steam shovels of the railroad type could swing their booms and dippers through an arc of only about 200 degrees. So even when fitted with a long boom, a railroad shovel was severely limited when used as a stripping shovel.

When the big fully revolving shovels came into popular use after 1910, they were also mounted on rails, but they required two sets of parallel railroad tracks to support their weight. Stripping shovels first traveled on crawler tracks in the early 1920s. Crawlers were slow to be adopted at first because of their much higher cost, about 40 percent higher than the traditional rail mountings. But when one or two pioneering owners found that the extra cost was soon offset by the shovel's superior mobility, the elimination of the rails, and the reduction of the ground crew from about eight workers to one, crawler mountings became very popular and orders for rail-mounted shovels diminished.

The Bucyrus 80-B was the first stripping shovel to be equipped with crawler tracks. Weighing a mere 81 tons, and carrying a dipper of 2-1/2 yards, it was small by stripping shovel standards. When first offered in 1922, however, it was the largest crawler-mounted shovel yet constructed by any manufacturer. Instead of the usual railroad mounting, the 80-B was equipped with four crawler tracks, an innovative design for its day. It was somewhat of a hybrid machine, as

A close-up view of the four-crawler undercarriage of the Rapier 5160. Each crawler is mounted under a vertical hydraulic cylinder, and the machine was leveled by adjusting the hydraulic flow to each cylinder by means of the valve wheels seen at left.
—Keith Haddock

A Rapier 5365 built in 1940 is shown stripping ironstone in central England for Stewarts & Lloyds Minerals Ltd. The 700-ton machine carried a 9-yard dipper. Note the iron beam attached to the boom to serve as an overhead crane for maintenance.

—Keith Haddock

most 80-Bs were equipped with regular short booms for loading hoppers or rail cars. Because it was not a truly dedicated stripping shovel, it is not included in the stripping shovel table in Appendix 3.

The much larger 320-B, which first appeared in 1924 as a 7-1/2-yard rail-mounted machine, became the first true stripping shovel fitted with crawler tracks in 1925. Bucyrus replaced the rail mountings with eight crawler tracks mounted in pairs at each corner. Marion quickly responded by placing crawler tracks on its record-size Model 350, which had an 8-yard dipper. The eight-crawler mounting pioneered in the 1920s became the standard configuration for all the giant stripping shovels in later decades.

The crawler tracks of the eight-crawler undercarriage, or lower frame, are mounted in pairs at each corner of the machine. Each two-crawler assembly is attached to a large vertical hydraulic cylinder (or screw arrangement, in the case of some early machines), which ensures that the machine can be kept level when working on uneven terrain. This leveling device is particularly important when the machine is propelling because it allows the crawler assemblies to move vertically when negotiating uneven ground. The arrangement minimizes stresses in the lower frame. The leveling is usually done automatically by means of a pendulum system that directs the hydraulic fluid to and from the four cylinders to keep the machine on an even keel. Each crawler assembly is steered by horizontal hydraulic rams that move the front pair, or the rear pair, in unison.

Since the late 1920s, stripping shovels have been provided with independent propel motors for each crawler assembly. On the larger machines, eight powerful motors are installed, one to drive each crawler

A rare photo of the only stripping shovel built by the German company, Menck & Hambrock. Built in 1927, its 8-1/2-yard dipper and 540-ton operating weight made the four-crawler machine one of the largest shovels in the world at that time.

—Heinz-Herbert Cohrs collection

track. Even with all this power, the top speed of a stripping shovel is only about a quarter-mile per hour. That may not seem like much, but a shovel's worth is measured by how much material it can move in an hour, not how fast it can travel. In fact, when a shovel is moving, it's not making any money for its owners.

POWER FOR THE BIG STRIPPERS

Up until the mid-1920s, stripping shovels were powered by electricity or steam. Since then, they have been exclusively electric-powered. The power for stripping shovels made the jump from steam to electric without the typical diesel phase seen in other types of equipment. The reason for this is that steam engines or electric motors can be fitted on the machine exactly where power is needed, and avoid long-distance transmission of power via gear trains, long shafts, or chains from one single diesel engine. Imagine having a chain or a driveshaft long enough to power a shipper shaft halfway up a boom measuring over 200 feet long!

In 1919 both Marion and Bucyrus applied an improved control system to their electric machines—Ward Leonard control. It produces precise variable speed similar to steam power and proved well suited to excavators. Ward Leonard uses AC-powered motor generator sets to power DC motors for the machine's motions. With Ward Leonard control, maximum pull is available at motor stall speed, and a motor will not burn out when overloaded. Bucyrus first applied Ward Leonard control on a 225-B stripping shovel, while Marion put it on the Model 300-E. Since then, Ward Leonard control has been standard on all stripping shovels.

The magnificent lines of the Marion 5323 stripper at work for Marquette Cement near Superior, Ohio. This particular machine was ordered with a special 160-foot-long boom, whichreduced the dipper capacity from the standard 17 yards to 11 yards. It featured Marion's standard knee-action crowd driven by rack and pinion from the gantry.

—Keith Haddock collection

OPERATION

The operation of a large stripping shovel takes careful planning. Because of the high ground pressure under its crawlers, the stripping shovel must always operate on a very stable footing, such as a coal seam. The coal the shovel is uncovering makes an ideal base, but if the shovel runs off solid material, it can quickly sink and bog down. This must be avoided at all costs, because when the biggest machine on the site is stuck, nothing can pull it out! It may take scrapers and bulldozers several days to excavate around the large shovel and haul in dry material before the big machine can be driven out under its own power.

Pit wall failure is another hazard to avoid. Sometimes the pit wall where the shovel is working becomes unstable and large chunks of rock come crashing down, partly burying the machine. A wall failure can occur without warning, and the stripping shovel's slow propelling speed makes a quick escape impossible.

Planning a mining operation with a stripping shovel includes breaking in a new cut, or ramping the machine down from ground elevation to the pit floor level. This can be a tricky operation, as the machine must travel down a ramp designed as steep as possible to minimize the amount of material to construct it. In many cases, the ramp is dug with other equipment, so the shovel can commence digging at its most-efficient face height.

The operating crew of a stripping shovel consists of three or four people, depending on the size of the

The second-largest shovel built by Marion was this 5960, owned by Peabody Coal Company. Known as the *Big Digger*, the 125-yard machine weighed 7,338 tons. It started work in 1969 at the River Queen Mine, near Central City, Kentucky. In 1989 the mine closed and the machine was scrapped.

—Keith Haddock

machine. The operator, oiler, and ground man are joined by an electrician on some of the larger machines. The oiler's duties, in addition to the obvious one, include keeping the machine clean. He also functions as a relief operator, and spends time in the operator's seat to become fully familiar with all the machine's operations. The oiler is usually "second in command," and next in line as operator when fully trained. The ground man takes care of the lower works of the machine. He ensures the power cable is kept away from falling rocks and out of the way of other vehicles. He also operates a bulldozer usually assigned to work with the shovel. The bulldozer's job is to keep the working area level, push back errant rocks that roll close to the shovel's tracks, and construct roads for maintenance vehicles to access the shovel.

The operator controls the machine's digging motions with two hand levers and two foot pedals. The right lever controls the hoisting of the dipper, and the left lever controls the crowd action (handle extension and retraction). The two foot pedals control the swinging of the machine, right or left—one pedal for each direction. The levers and pedals are arranged so that in their central positions, power is off. As a lever or pedal is moved off its central position, power is applied through the DC motors to the selected motion—hoist, crowd, etc. As the lever or pedal is moved further, power is stepped progressively to maximum. When the lever or pedal is moved back in the opposite direction, power to the motors is reversed so that the dipper moves in the opposite direction. It all sounds very simple, but because no brakes are used during the digging cycle, the operator needs a lot of skill to synchronize the hoist, crowd, and swing motions. And the machine's motions don't change instantly. Even though the dipper and handle appear to move slowly, their massive weight takes several seconds to come to a stop or move in the opposite direction. The operator must therefore anticipate the dipper's movements well in advance.

In addition to the four main controls mentioned above, there are many other switch controls in the operator's cab. "Parking" brakes for the hoist and crowd machinery, for example, are not used during the digging cycle, but only for holding the drives stationary when power is cut off. Other switches control the many lights, air conditioners, motor excitation, and emergency stop.

Steering and propelling are usually controlled by the ground man from a sub cab in the lower frame. From here, he controls the direction of travel and the hydraulic steering rams connected to each crawler assembly. On stripping shovels built since the 1920s, four vertical hydraulic leveling rams level the machine automatically.

Because of their high capital cost, stripping shovels work around the clock, seven days a week, to gain maximum returns for their owners. Although they may shut down for up to 11 statutory holidays a year, they can still clock as many as 6,000 annual operating hours, depending on their availability and the demand for the mineral being produced.

Most stripping shovels cycle every 50 to 80 seconds. This may seem slow, but when you consider the huge amount of material being scooped up each cycle, hour after hour, and the great distance it is moved, you quickly realize what vast quantities of material these shovels can move. Stripping shovels can dig harder material than a similar-sized walking dragline (chapter 4), but drilling and blasting are still recommended on most types of overburden to gain maximum productivity from the machine, and to prolong its life.

MOVING THE BIG MACHINES

These giant stripping machines are so large that they have to be shipped in many pieces and erected at the site where they operate. When the machine is new, all parts are shipped directly from the factory to the site and erected under the supervision of the manufacturer's representative. Parts are shipped by rail or road or both. A new machine can take more than 250 truck or railcar loads of parts to complete delivery, and erection time can be up to 1-1/2 years.

A stripping shovel may work its entire life at one location, and encounter both its erection cranes and scrapper's torch at the same site. In other cases, it may be moved to a new location because reserves have run out, or it could be sold to a new owner. When it comes time to move these monster shovels, they can't simply be loaded onto a flatbed and hauled down the road. They must be broken down into pieces as large as possible, while still suitable for transport on highway trucks or railcars.

Moving a machine of this size takes a crew with special skill and a lot of experience. The original manufacturer is usually involved in such moves, and may provide a skilled supervisor to make sure every phase of the operation is carried out safely, and in accordance with the manufacturer's recommendations. Sometimes major components are simply unbolted to disassemble them, but usually the cutting torch is used, and parts rewelded during assembly. The manufacturer's representative advises how the machine should be broken down into components, and how it can be reassembled so the machine can go back to work without problems.

When a stripping shovel is erected at a new site, it is actually built in the pit where it will start to dig. The pit is dug beforehand by other equipment, and the parts are assembled on a specially prepared base below ground level. This method eliminates the need for the shovel to ramp itself into the pit, and cranes used in the erection can be smaller, since components are positioned at or below ground level. The cranes don't have to lift materials as high, and their operating radius can be kept small.

The shovel's erection site must be laid out in an orderly fashion. When dismantling and reerecting a machine, careful planning is required. The first component coming off the machine being disassembled is one of the last pieces needed for reassembly—because it is disassembled from the top down, and reassembled from the ground up. Thus, there must be lots of storage space for components to be laid on blocks, railroad ties, or clean gravel. When a machine is moved it is usually a good time to repaint, and additional space should be arranged so most of the large components can be primer-painted on the ground if possible. Extra space is also needed so that some subcomponents can be preassembled on the ground, and then lifted into place by a crane. One large derrick crane supplemented by several rubber-tired mobile cranes is typically the preferred crane fleet on large excavator erection sites.

CHAPTER TWO

STRIPPING SHOVEL CHRONOLOGY

Coal mining contractor Wright & Wallace used a land dredge in 1885 to strip coal at Mission Field near Danville, Illinois. This was the earliest recorded application of a digging machine used in strip mining. The contractor bought an old dredge made of wood, removed the hull, and placed the machine on rollers. Workmen installed a wooden boom 50 feet long and a dipper of 1-1/4-cubic yard-capacity. Several similar land dredges were made before the turn of the century, but these wooden machines had very short working lives.

Bucyrus-Erie's super shovel in the 1950s was the 1650-B. The first one, carrying a dipper of 55 yards, went to Peabody Coal Company in 1956, and was named the *River Queen* after the mine in Kentucky where it worked. After some 13 years at River Queen, the 1650-B saw action at three more mines in Kentucky—two for Peabody, and a third when it was sold to Green Coal Company to work at its Henderson County Mine.

—Bucyrus International Inc.

The earliest known revolving stripping shovel was this rail-mounted machine built by John H. Wilson & Company in 1900. Used in England to uncover iron ore by Lloyds Ironstone Company, it had a 1-1/2-cubic-yard dipper on a boom 70 feet long. It weighed 78 tons, and had a working life of 54 years!
—Keith Haddock collection

In 1899 the well-established Vulcan Steam Shovel Company built two long-boom stripping shovels to mine phosphate for the Berkley Chemical Company. These successful machines were known as "Vulcan Phosphate Specials," and seven more were built in 1900. Vulcan continued to build long-boom shovels for stripping applications. In 1907 a surface coal mining company installed a long boom on a Vulcan Model K railroad-type shovel and used it to strip overburden near Lily, Kentucky.

The largest Vulcan stripper went to work near Tulsa, Oklahoma, in 1910. This was the Class L, a 130-ton, 2-cubic-yard land dredge with a 60-foot boom. It was profitably used to strip 17 feet of overburden from a coal seam only 2 feet thick. Later that same year, the Vulcan Steam Shovel Company and its designs were purchased by the Bucyrus Company.

All of the above Vulcan machines and the earlier dredges were of the nonrevolving type. They were designed on the railroad shovel principle, in which the boom and dipper arm were capable of swinging only about 200 degrees instead of a full 360 degrees. Although profitable, these part-swing machines were severely limited in their range. Operators wanted a wider cut with adequate room to load the coal, and the capability to dump material anywhere within the machine's radius. A fully revolving shovel was the obvious answer. Not only would it dig a wider cut, it could also be used as a crane to relocate sections of the shovel's own rail track from behind to in front of the machine.

FIRST FULLY REVOLVING STRIPPING SHOVELS

The first stripping shovel, as we know it today, was a fully revolving machine built in 1900 by John H. Wilson & Co. in England. Operated by Lloyds Ironstone Company to uncover iron ore, it carried a dipper of 1-1/2-cubic yards capacity on a boom 70 feet long that provided a 60-foot dumping radius. The 78-ton unit was mounted on rail tracks, and had a wire-rope swing system. It was regarded as a marvel in its day, when most excavation work was done by hand. So successful was this first stripper that it had a working life of 54 years!

Another early full-circle stripping shovel was the Grossmith of 1908. Used in England for mining brick clay, this giant machine carried a 3-yard dipper on a boom 80 feet long. The boom point sheave was 8 feet in diameter, and the swing circle rail was 18 feet in diameter. This machine was fitted with a unique power-operated variable-pitch dipper device. Driven by a steam turbine motor, this device allowed the dipper to be adjusted relative to the handle, giving the shovel a greater floor-cleaning radius. But the device did not survive the jolting of the dipper, and was soon discarded. The variable-pitch dipper did reappear on a few large shovels in the 1960s through the 1980s.

Another early revolving stripping shovel was this Grossmith, built in England in 1908. It carried a 3-yard dipper on a boom 80 feet long. In 1926 the machine shown here was converted from steam to electric power. It was initially equipped with a steam-operated bucket tilt mechanism.

—Keith Haddock collection

The Bucyrus Company built the first full-circle long-boom stripping shovel in the United States soon after it purchased the Vulcan Steam Shovel Company in 1910. Known as the Class 5, and built to Vulcan designs, the fully revolving shovel carried a 1-1/2-cubic-yard dipper on a 55-foot boom. A total of three Class 5s were shipped to customers in 1910–1911, and all worked on coal stripping in the Pittsburg, Kansas, area. These machines worked well for their owners, attracting much attention from the industry and other manufacturers.

The first long-boom stripping shovel from the Marion Steam Shovel Company went to work in 1911 at Mission Field near Danville, Illinois. At first, Marion was very reluctant to build the machine, stating it "could not guarantee its operation because of its large dimensions." Modern welding had not been developed—riveted construction was the order of the day. But the customer, Grant Holmes and W. G.

Below: The First evolving stripping shovel in the United States was the Marion 250 of 3-1/2 cubic yards capacity. Purchased by Mission Mining Company in 1911, it was operated near Danville, Illinois. Another early user was T. J. Forschner Coal Company. Its machine, shown here, was one of 19 ordered by coal companies between 1911 and 1913.

Hartshorn, convinced Marion to build the machine. Known as the Model 250 and built on solid, heavy-duty designs, it was a resounding success. This machine set the stage for large-scale mechanized stripping of minerals by surface mining methods.

The Marion 250 was a steam-powered, rail-mounted machine, having a working weight of 150 tons. It carried a 3-1/2-cubic-yard dipper on a 65-foot boom. Even at this early date, Marion had pioneered the concept of utilizing a hydraulic cylinder under each corner for equalizing stresses and leveling the machine. This concept has been used in every Marion stripping shovel. When the 250 was introduced, the industry was beginning to recognize the advantages and potential of being able to strip away the entire depth of overburden with a single, large machine. The Marion 250 was so successful that, by 1913, the company had 19 orders.

Hot on the heels of the 250, Marion introduced an even larger shovel, the 5-yard Model 270 in 1912. This was updated to the heavier but same-capacity Model 271 the following year. The 271 briefly held the title of the world's largest operating shovel.

Just as the first Marion 250 went to work, Marion's chief competitor, Bucyrus Company (later Bucyrus-Erie Company, and now Bucyrus International, Inc.), thought it had secured an order for a railroad shovel and a dragline to strip coal in tandem. The tandem operation was a common way to strip coal prior to the introduction of the stripping shovel. But this required the material to be handled twice: the shovel dug it from the face, while the dragline cast the same material clear of the excavation. When

The Bucyrus Company worked rapidly to produce a stripping shovel to compete with Marion's 250. The result was the Model 150-B, appearing in 1912, and shown here operating for C. F. Markham Coal Company at Fuller, Kansas. The shovel was leveled by a screw jack system, and carried a 2-1/2-cubic-yard dipper.
—Bucyrus International Inc.

Bucyrus' customer went to observe the Marion machine in operation, the company promptly canceled the order in favor of the single Marion machine.

Response from Bucyrus was immediate. Its engineering department went to work, and within a few months, came out with competitive machines, the 2-1/2-yard 150-B and the 3-1/2-yard 175-B stripping shovels. These were also rail-mounted steam machines, but instead of using hydraulic jacks, Bucyrus opted for a three-point suspension system with screw jacks for leveling. By the end of 1912, one of each of these shovels was already at work, stripping coal in the southeast Kansas coal field.

SHOVEL SIZE INCREASES

The idea of stripping coal with large shovels caught on very quickly, spurred on by the demands of World War I. During this period, the only two manufacturing companies in this business, Bucyrus and Marion, worked feverishly to introduce new models and capture the world record for size. In 1914, Bucyrus designed a new monster shovel for its day, the 225-B. It weighed more than 300 tons and carried a 6-yard dipper on a boom 75 feet long. The first machine was shipped to Carney-Cherokee Coal

Company for work at Mulberry, Kansas. The 225-B was a great success, and more than 90 of these huge machines were erected until the last in 1923. This turned out to be the Number One best-selling stripping shovel. Marion followed suit with its Model 300 in 1915. This had the same capacity as the 225-B but weighed some 30 tons more and carried a boom 15 feet longer.

In 1923, Marion broke the world size record with the Model 350, a huge shovel weighing no less than 560 tons! This giant had an 8-yard dipper on a 90-foot boom and was the world's largest mobile land machine of any type ever constructed up to that time. Initially rail-mounted, the 350s were equipped with crawler tracks after 1925. The same machine was also offered as a dragline, with the model designation changed to the 360. These machines were available with either steam or electric power.

The drive to the crawler tracks on the 350 came from the main hoist motor upstairs in the revolving frame. A long train of bevel gears, shafts, and spur gears brought the power down through the center pintle of the machine and then out to each of the four crawler assemblies through a series of jaw clutches. A further set of jaw clutches split the drive to power the steering of each crawler assembly through screwed rods. The front pair of crawlers could be steered independently of the rear pair so the machine could negotiate a relatively sharp curve, or "crab" if all crawlers were steered in the same direction. A two-speed travel was also incorporated so the inner crawlers could be propelled in low gear and the outer ones in high gear to aid turning corners.

To propel the machine, the operator in the cab had control of the main hoist/propel motor but no control over steering. The ground man had to run around under the machine, heaving the heavy jaw clutches, and sliding the gears in or out to effect steering. He carried a heavy iron tube which he placed over the levers to obtain greater pull. Another duty of the ground man was to level the machine. Marion used its patented four-cylinder hydraulic leveling device, but leveling the hydraulic jacks at each corner of the machine was not automatic. Guided by two pendulums attached to the frame, the ground man operated valves to divert hydraulic pressure from one cylinder to another until the machine was level. Such large-scale use of hydraulics on a machine of this size in the 1920s is truly amazing.

Imagine the sight of this monster machine eating up the countryside, taking 8 cubic yards at each bite

A new monster shovel appeared on the scene in 1914. It was the 225-B, wielding a 6-yard dipper, weighing more than 300 tons, and offered with steam or electric power. This one is digging a river diversion channel for the Hydro Electric Power Commission near Niagara Falls, Ontario. The 225-B turned out to be the all-time best-selling stripping shovel. More than 90 were sold up to 1924. —Bucyrus International

Marion brought out its Model 300 stripping shovel in 1915 to compete in the same size range as its competitor's 225-B Model. This standard rail-mounted, steam-powered shovel is shown uncovering coal for Elsworth Klaner Construction Company near Pittsburg, Kansas. —Keith Haddock collection

Above:
The Marion 350 was the world's largest mobile land machine when it appeared in 1923. A 560-ton roving monster was certainly an eye-opener in the days of the Model T Ford and silent movies! Forty-seven of these giants, including the 360 dragline version, were built up to 1929. The last remaining Marion 360 was converted to a shovel and has been preserved at the Diplomat Mine Museum near Forestburg, Alberta, Canada.

—Keith Haddock

Left:
A view showing the massive crawler drive arrangement under the Marion 350/360 shovels. Two driveshafts run to each of the four crawler assemblies—one to propel the crawlers, the other to facilitate steering through a screwed rod arrangement. All power is transmitted through the center pin from the hoist motor in the upper works.

—Keith Haddock

and dumping its load half a football field away. The massive proportions of this machine sharply contrasted with the automobiles of the era, with their narrow wheels and light chassis. And all this was at a time before sound movies were in the theaters—even before the era of welding!

The 350 series machines were a great success, with 47 units sold before production ended in 1929. The last one to operate finally shut down in 1980 at the Diplomat Mine near Forestburg, Alberta, Canada. It is now preserved in the Diplomat Mine Museum on the site where it once operated.

Two competitive machines appeared on the scene in 1924, the year after Marion introduced its 350. These were the Bucyrus 320-B and the No. 300, made in England by Ruston & Hornsby Ltd. They were not quite as large as the record-beating Marion, but were significant machines nonetheless. They were both in the 8-yard class, and weighed in at about 390 tons. Initially steam driven and rail mounted, crawler tracks and electric power were offered on later models. The 320-B proved to be a great success, and 29 shovels plus 8 dragline versions were sold before the last one was shipped in 1930.

The steam versions of these three giant shovels of the 1920s—the Marion 350, the Bucyrus 320-B, and

The Bucyrus 320-B was introduced in 1924 as a 7-1/2-yard, rail-mounted steam shovel. Later, crawler tracks were added, making this the first stripper to be so propelled. The rail-mounted version shown here is removing overburden near Pittsburg, Kansas, for the Litchfield Coal Company.

—Bucyrus International

the Ruston No. 300—were the largest steam-powered shovels ever built. In their time they provided one of the most impressive sights and sounds of the mechanical world. The unique hissing and snorting of the main hoisting engine on the deck, combined with the separate swing engines, and the hissing of the crowd engine mounted on the boom must have created an unforgettable symphony, but something that was probably not appreciated by those working around the machine! As shovel dipper sizes continued to spiral above 8 cubic yards, all machines changed to the relatively silent, electrically operated variety.

Both Marion and Bucyrus introduced stripping shovels of 12 cubic yards capacity in 1927. Marion debuted its 5480, incorporating features from its successful Model 350 and weighing almost 1,000 tons. Bucyrus launched its 750-B, the forerunner of a long line of giant stripping shovels produced by that company.

Another contender in the 8-yard class was the Ruston 300, introduced in 1924 by the British firm, Ruston & Hornsby Ltd. Like its competitors, it was available with steam or electric power, and later models were offered with crawler mountings. The steam versions of the three giant shovels—the Marion 350, the Bucyrus 320-B, and the Ruston 300—were the largest steam-powered shovels ever built.

—Keith Haddock collection

Ten 750-Bs were delivered up to 1930. Then the Michigan Limestone Company of Rogers City, Michigan, asked Bucyrus-Erie Company if it could design a more efficient 750-B. The result was the first shovel to be fitted with a counterbalanced hoist system. Now the 750-B's capacity could be increased 50 percent, to 18 cubic yards, and more importantly, the machine actually used less power. The first of this new breed was delivered to Michigan Limestone in 1930.

The counterbalanced hoist consisted of a moving counterweight at the rear of the revolving frame and connected to the dipper hoist drum through its own set of wire ropes. The counterweight ran vertically in a cage similar to an elevator, balancing the weight of the empty dipper. Thus all the hoisting power could be applied to filling the dipper and hoisting the load. Three more 750-Bs with counterbalance hoist system were delivered, including the last in 1940. These included a second machine for the Michigan Limestone Company, and by that time its capacity had increased to 22 cubic yards.

Marion had a counterbalanced hoist on some of its 18-yard Model 5560 shovels that appeared in 1932. But Marion opted for a rack-and-pinion system to raise and lower the counterweight. A wire rope ran from the hoist drum to a drum shaft mounted on the moving counterweight. This shaft had a pinion at each end that meshed with vertical racks attached to a frame at the back of the machine. Guides held the massive weight in place as it moved up and down in harmony with the dipper.

In 1929, Marion claimed title to the world's largest land machine again, when the 1,500-ton Model 5600 was sold to United Electric Coal Companies for use in southern Illinois. Although only one was ever built, the 5600 had a very interesting and long life in various configurations. Initially the machine was purchased as a 15-yard shovel. After only four years' work, the owners converted the machine to a 20-yard dragline, the largest dragline ever to work at that time. Then in 1937 the 5600 was converted back to a shovel. But this time, the dipper was increased to 26 cubic yards. With the advent of World War II, the machine sat idle for many years. In 1957 the 5600 stripper was converted into a bucket wheel excavator, and with this continuous excavator make-over, the output of the machine was increased five-fold over the original shovel's production. The wheel excavator, with its original 5600 base, continued in operation until 1984. It was finally scrapped in 1994.

Now it was Bucyrus-Erie's turn to make a move on the world's largest machine title, and this it did handsomely in 1935 with the innovative 950-B. This 30-yard stripper introduced many features that have

Weighing almost 1,000 tons, the 12-yard capacity 5480 stripping shovel was launched by Marion in 1927, nudging shovel capacity ever upward. It came equipped with crawler tracks as standard. Reaching high is this 5480 in Missouri, working for Minden Coal Company. Note the trailing power cable wrapped on the reel, and the electric crowd motor on the boom.
—Bucyrus International Inc.

remained in Bucyrus shovel design ever since. The most notable feature was the front end arrangement, with a single tubular dipper handle operated by wire ropes. The handle's tubular design allowed it to rotate, thus minimizing stresses on the handle and boom caused by unbalanced loads on the dipper. The rope-operated crowd motion allowed the crowd machinery and motor to be mounted on the revolving frame, where it minimized swing inertia and saved energy. The two-part boom was pinned at its center and tied back to the gantry with two heavy beams. This eliminated all bending stresses at the center of the boom, and with the center of the boom tied back to the gantry, the entire boom could be of much lighter construction.

The lower works of the 950-B also received radical improvements. A propelling motor was mounted in each of the four crawler assemblies, eliminating the previous complicated system of multiple gear trains, shafts, and jaw clutches. And for the first time on a Bucyrus-Erie machine, a hydraulic leveling system was introduced, dispensing with the earlier screw-types. This system automatically leveled the machine through a pendulum system. Steering was also effected through large horizontal hydraulic cylinders. The 950-B had an operating weight of some 1,250 tons.

It wasn't many months before Marion produced a stripping shovel to compete with the 950-B. It was an upgraded version of the previously available 5560, but its capacity was boosted to 32 cubic yards, made possible by using high-tensile steel in the front end, and more efficient electric motors. Its weight was increased to 1,550 tons, some 345 tons heavier than the previous 5560.

The following is an excerpt from *Pacific Road Builder & Engineering Review*, December 1935:

> The first of the new giant Bucyrus-Erie 950-B power shovels has recently been put into service by

This Bucyrus 750-B, also launched in 1927, competed in the same size range as Marion's 5480. The later versions, like the one shown here, were equipped with a counterbalanced hoist system (elevator-like structure at the back of the house), which increased the dipper capacity from 12 to 18 cubic yards.
—Keith Haddock

the Binkley Coal Company of Indiana on its coal stripping operations near Terre Haute. Some idea of the capacity of these mammoth excavators may be had by comparison with the shovels that dug the Panama Canal. Three of the new 950-B machines could move the same amount of dirt in the same time that it took 90 of the then-largest excavators to dig the 'big ditch.'

Nor is this new giant digger large only in dipper capacity. In operations, the huge 30-yard dipper lifts 45 tons of dirt and rock at each pass, to be dumped farther and higher than any shovel in the world has ever been able to dump. The mammoth machine has a boom 105 feet long, a dipper stick 64 feet long, a maximum dumping height of 70 feet, and a dumping radius of 106 feet. Speed is another attribute in which the 950-B excels—never before has there been offered a stripper with a digging cycle as fast as that of this revolutionary new shovel. On large stripping open pit mining projects, this means tremendous savings in costs.

The operator of the machine sits in his control cab at the height of a three-story building. At his finger tips are the controls for 32 different electric motors used in the various operations of moving, leveling and digging. The new 950-B is more than a third larger in dipper capacity than the 750-B, the previous largest stripper, and combines this with increased range, speed and convenience.

RAPIER STRIPPING SHOVELS

The established stripping shovel leaders had some competition from England in the 1930s. The old engineering firm of Ransomes & Rapier Ltd. of Ipswich, England, sold its first steam shovel in 1914. Then in 1924 the company entered into an agreement with Marion Steam Shovel Company to build certain Marion-designed excavators under license in the United Kingdom. The first Rapier stripping shovel went to work in 1934 stripping iron ore for Stewarts & Lloyds Ltd. In England. The machine was the 5360, based on the former Marion 350 shovel. It was a resounding success, but by the time the second of these machines was delivered in 1936, the Rapier-Marion agreement had terminated. Over the next eight years, however, Rapier went on to build 13 of these giant stripping shovels, as well as some smaller ones (see table in this chapter).

Each 5360 series machine was modified from the original Marion design and tailor-made for its operating location. And since the Marion agreement had terminated, the "Rapier" content gradually increased. Each machine with major modifications was recognized with a new model designation, such as 5361 and 5362. The Rapier 5360 series machines were somewhat larger than the former Marion 350s. Most of them had booms 104 feet long, and

Like Bucyrus, Marion incorporated a counterbalanced hoist system on some of its 18-yard Model 5560 shovels, which first appeared in 1932. The first 5560 is shown still working in 1980 for Clemens Coal Company, Pittsburg, Kansas.

—Keith Haddock

dippers from 9 to 11 cubic yards. They were all rail-mounted on twin rail tracks, and electrically operated. World War II brought stripping shovel production to an end, and the last one, a 5367, went to work in 1943. Rapier later became famous for its giant walking draglines, but the company built no more stripping shovels.

Rapier made plans to reenter the stripping shovel business in 1963. It designed the S1000 (20 cubic yards) and the S1500 (30 cubic yards) machines, but never built them. Adverse marketing conditions and a change in company policy forced cancellation of the new shovels, as well as a temporary withdrawal from the walking dragline business. (See Appendix 1 for a brief history of Ransomes & Rapier Ltd.)

KNEE-ACTION CROWD

Marion's knee-action crowd in 1940 was the next major design breakthrough in stripping shovels. It featured a totally new design of front-end geometry. Instead of attaching the dipper handle to the boom, the handle was attached to a stiff leg that pivoted at the boom foot. The crowd machinery was mounted on top of the house and operated through rack and pinion or wire ropes. The Model 5561 was first to receive the new front end, which, with its 35 cubic yards capacity, claimed a dipper size record yet again. Nicknamed the "grasshopper" front because the long handle and stiff leg appeared to fold like the insect's leg, the knee-action crowd became standard for all subsequent Marion stripping shovels, and later, the Bucyrus 1950-B. By 1945 the 5561 had grown to 40 cubic yards in capacity.

> **ADVANTAGES OF THE KNEE-ACTION CROWD**
> 1. The crowd machinery is moved to the gantry, near the machine's center of rotation, so swinging inertia is minimized.
> 2. Torsional and bending stresses are removed from the boom, because the dipper handle is connected to a movable stiff leg instead of the boom. Thus the boom can be made considerably lighter.
> 3. The action of the movable stiff leg allows the dipper to move in a long horizontal sweep at ground level. This results in a long clean-up radius and positions the dipper teeth so that they will not gouge into the coal.
> 4. The crowd motion is capable of producing a tremendous downward force at the toe of the bank, just where it is needed in hard digging.

Bucyrus-Erie's response to Marion's 5561 was the 1050-B. It was an expanded version of the earlier 950-B, and first appeared in 1941 as a 33-yard machine. It retained the two-part boom and rope crowd design. A total of 12 1050-Bs were shipped until the last, a 45-yard machine, went to United Electric Coal Company's Banner Mine, Illinois, in 1960. At the time of writing, this machine is one of the last stripping shovels still operating. It is now at the Industry Mine, Illinois, of Freeman-United Coal Mining Company.

RAPIER STRIPPING SHOVELS

Model	Serial No.	Owner	Location	Boom (Ft.)	Dipper (CuYd)	Year Ordered	Working From	Working To	Year Scrapped
5160	401	Appleby-Frodingham	Cringle, Lincs.	55	4	1936	1936*	1960	1961
5160	851	Alpha Cement Ltd.	Shipton, Oxford	57	3-1/2	1938	1938	c. 1978	c. 1980
5360	192	Stewarts & Lloyds	Rockingham, Corby	93.5	9	1934	1934	1968	1969
5361	284	Stewarts & Lloyds	Deene, Corby	104	11	1935	1936	1952	
			Market Overton				1952	1971	1973
5360	440	S. Durham Steel & Iron Co.	Irchester	104	8	1936	1937	1969	1969
5362	502	Stewarts & Lloyds	Earlestrees, Corby	93.5	9	1936	1938	1973	1974
5360	509	Appleby-Frodingham	Colsterworth, Lincs.	104	8	1936	1938	1968	1968
5360	669	Richard Thomas & Baldwins	Finedon	104	9	1937	1939	1954	
			Blisworth				1954	1968	
			Buckminster				1968	1972	1972
5363	963	Stewarts & Lloyds	Cowthick, Corby	104	9	1939	1940	1960	
			Sibleys, Gretton				1960	1976	1981
5364	1016	Stewarts & Lloyds	Park Lodge, Gretton	104	9	1940	1940	1976	1976
5365	1017	Stewarts & Lloyds	Corby	104	9	1940	1941	1964	
			Harlaxton				1964	1974	
			Park Lodge, Gretton				1975	1980	1981
5360	1079	Park Gate Iron & Steel Co.	Sproxton, Lincs.	104	8	1940	1941	1973	1974
5360	1117	Appleby-Frodingham	Colsterworth, Lincs.	104	8	1940	1941	1962	1962
5366	1163	Stanton Iron Works	Geddington, Glendon	104	9	1940	1942	1979	1981
5367	1164	Cranford Ironstone Co.	Cranford South	104	9	1940	1943	1969	1978

* Dragline from 1946

ERA OF THE SUPER STRIPPERS

The boom in coal-fired power generation in the 1950s produced a demand for even larger, more efficient stripping machines. The Hanna Coal Company (now Consolidation Coal Company) ordered a shovel of unbelievable proportions for its operations in eastern Ohio. Much larger than anything before, the *Mountaineer* took its first bite in January 1956, and the era of giant stripping shovels was born. The Marion-built monster, the Model 5760, wielded a 65-cubic-yard dipper, and could tackle a face over 100 feet high. A detailed story of the *Mountaineer* appears in chapter 3. As amazing as the *Mountaineer* appeared to the onlookers at its launching ceremony, few would have suspected that one short decade later, the same manufacturer would launch a shovel three times its size.

After the *Mountaineer*, the size of stripping shovels increased astronomically. Almost immediately,

When Bucyrus-Erie launched its 950-B stripping shovel in 1935, it marked a new era in shovel design. The huge 30-yard machine incorporated many new features that are still found in modern machines—like the tubular dipper handle and hydraulic steering. Pictured is the 950-B *Mr. Diplomat* located at Forestburg Collieries' Diplomat Mine.
—Keith Haddock collection

A big breakthrough in stripping shovel design occurred in 1940, when Marion introduced the knee-action crowd on the new 35-yard Model 5561. The machine pictured is working at the Power Mine of Peabody Coal Company, Missouri, in 1980.
—Keith Haddock

Bucyrus-Erie responded with a shovel of similar size, the 55-yard 1650-B *River Queen*, ordered by Peabody Coal Company for its River Queen Mine in Kentucky. Peabody also ordered a Marion 5760 for its River King Mine in Illinois in 1957, but this time the dipper was increased to 70 cubic yards. Three more 5760s were built before it was upgraded to the 3,700-ton 5761 in 1959. With a 65-yard dipper and a boom some 20 feet longer than the model it superseded, the first 5761 captured the title of the world's largest shovel. It was delivered to Peabody's Lynnville Mine in Indiana.

Next came a huge jump in shovel size. Bucyrus-Erie took the spotlight in 1960 by announcing that it had secured an order for the world's largest shovel, a Model 3850-B. The 3850's 115 cubic yards dipper capacity was almost double the size of largest shovel

Above: Bucyrus-Erie brought out its 1050-B in 1941 to compete with Marion's 5561. The machine was initially rated at 33 cubic yards capacity, but later versions were upgraded to 45 yards. At the time of writing, the 1050-B pictured is still operating at Freeman-United Coal Mining Company's Industry Mine, Illinois. This machine was purchased new in 1960.
—Urs Peyer

Right: In 1963, Bucyrus-Erie's 1850-B *Brutus* went to work for Pittsburg and Midway Coal Mining Company in Kansas. With its 90-yard dipper, it was the second largest shovel in operation at that time. Today the shovel is the focus of a mining interpretive center at West Mineral, Kansas. It is maintained by the preservation group, Big Brutus, Inc.
—Eric C. Orlemann collection

in operation at that time! Two of the 3850-Bs were built in quick succession, and delivered to Peabody Coal Company between 1962 and 1964. Their story is told in chapter 3.

Another super stripper was Bucyrus-Erie's 1850-B with a 90-yard dipper on a 150-foot boom. Christened *Brutus*, the 5,500-ton shovel went to work for Pittsburg & Midway Coal Mining Company at West Mineral, Kansas, in 1963. Standing 15 stories tall, the great machine required 150 railroad cars to deliver all its parts from the Bucyrus South Milwaukee factory. A crew of 52 men worked 11 months to put it all together. It was the second-largest stripping shovel to go to work at that time. After only 11 years work, this modern machine was idled, but not scrapped. It now stands high and proud on public display south of West Mineral, having been restored by Big Brutus, Inc., a nonprofit organization dedicated to the restoration and preservation of the shovel.

The race for the largest stripping shovel ended in 1965 when Marion broke the final record for shovel size. The incredible Marion 6360, named the *Captain*, was purchased by the Southwestern Illinois Coal Corporation to work at its Captain Mine near Percy, Illinois. With an operating weight estimated at 15,000 tons after additional modifications, this behemoth was truly the captain of all shovels. It was the largest machine of any type to move under its own

power on land, a record that has not been surpassed. The *Captain* is described in detail in chapter 3.

Although the race for larger and larger shovels had come to an end, the manufacturers didn't realize it. They expected the race to continue. They designed even larger machines, but there were no takers. Among others, Bucyrus-Erie designed the 4850-B at 220 yards, and then the 4950-B at 250 yards. This latter shovel would have had a weight of 18,000 tons supported on 16 crawlers, four per corner. There was even talk of a 1,000-yard shovel! But no more record-breaking shovels were built.

Although manufacturers would build no bigger machine than the Marion 6360, they would make several more stripping shovels before the last one in 1971. Bucyrus-Erie's final stripping shovel, a 130-yard 1950-B named the *Gem of Egypt*, was put to work in early 1967 by the Hanna Coal Company, now Consolidation Coal Company (Consol), in southeastern Ohio. This followed the first 1950-B, the *Silver Spade*, operated by Hanna in the same area since 1965. The *GEM* (Giant Earth Mover), as it later became known, was scrapped in the late 1980s. But the older of the two 1950-Bs, the *Silver Spade*, is still working at the time of writing. Read more about the 1950-Bs in chapter 3.

The last of the super strippers was the Marion 5900 sold to Amax Coal Company for work at its Leahy Mine, Illinois, in 1971. It was not only Marion's last, but the last by any manufacturer. It was one of two 5900s, and followed one sold to Peabody's Lynville Mine, Indiana, some three years earlier. The Leahy machine featured a variable-pitch dipper device that allowed the dipper to rotate relative to the handle. With this feature the shovel gained a much greater cleanup radius, and could uncover two seams of coal at different elevations at the same time. Through a valve, the operator controlled the flow of oil in a nonpowered hydraulic cylinder and cable arrangement. The shovel's own digging force changed the dipper angle as it moved from a flat position for cleanup, to its maximum tilt as it went up the steep digging face. The variable-pitch dipper idea was similar to that used on Marion's "Superfront" loading shovels from 1967; before that time an unsuccessful attempt was made with a 1908 Grossmith stripping shovel.

The Leahy Mine was absorbed into the adjacent Captain Mine in 1986, and the 5900 became part of the massive fleet of monster machines operating at Captain. It eventually replaced the Marion 6360 *Captain* shovel when it was destroyed by fire in September 1991. After the mine closed, the 5900 was scrapped in 1999.

DEMISE OF THE STRIPPING SHOVEL

In the right conditions, stripping shovels are one of the most economical methods of moving large volumes of earth. They have a faster average cycle time than walking draglines. They can also dig harder material than draglines. With all this going for them, it is perhaps surprising to find that the last of these monster machines was built as long ago as 1971. With the exception of a handful of two-crawler long-boom shovels purchased in the late 1970s and 1980s for special applications, the giant stripping shovel has grown obsolete.

Several factors contributed to the downfall of the stripping shovel. Most of the coal under relatively shallow overburden, ideal for stripping shovel operation, has now been worked out. Mines are moving into deeper overburden where stripping shovels are not as flexible as walking draglines. Working from the bottom of the pit, the stripping shovel's operating range is restricted to the dumping radius of its dipper. Today's new mines are designed for operation by walking draglines, which perform their work from the top of the face or at ground level, allowing freedom of movement around the pit to gain maximum reach. The dragline can even extend its working pad or bench by rehandling some of the material to make a "bridge" across the previously worked-out cut. It is then possible to operate the dragline from this backfilled bridge to gain even more reach. Because of this flexibility, a dragline can work in much deeper overburden than a shovel in the same size class.

When stripping shovels were king of the surface coal mines in the 1950s and 1960s, they were built in relatively small numbers. Markets for the coal were geared to the capacity of local power generating stations, which provided cheap, home-generated electric power. Once a mine was designed for a certain capacity, and the necessary stripping equipment installed, the status quo was usually maintained for several decades, during which time the equipment was not replaced. The very long working life of a typical stripping shovel is one reason why so few were built. Another reason is their colossal productivity. The giant stripping shovels served their coal markets handsomely, and uncovered huge amounts

of coal to meet the hungry appetites of the coal-fired generating stations.

Another reason for the demise of the stripping shovel is the rapid rise in coal output from the west during the 1980s. Particularly in Wyoming, coal is found in very thick seams and under very shallow overburden. Because western coal is cheaper to mine, and low in sulfur, much of this coal has captured the markets traditionally supplied by coal mines in the Midwest, particularly in the Illinois Basin, domain of the stripping shovels. Today, very few stripping shovels still operate, but the coal-fired plants still run at maximum capacity burning coal shipped from the west.

The last stripping shovel produced was this Marion 5900, sold in 1971 to Amax Coal Company's Leahy Mine in Illinois. It later worked at the adjacent Captain Mine, where it is shown operating with its 105-cubic-yard dipper. It features Marion's rope-operated knee-action crowd, and a variable pitch dipper system also found on Marion's Superfront shovels.

—Urs Peyer

CHAPTER THREE

THE SUPER STRIPPERS

Back in 1955 there were rumors of something big happening in the rolling hills of southeastern Ohio. Newspapers soon reported that a giant stripping machine of unbelievable proportions was being built in the area. The stories told of a monster shovel with a 65-cubic-yard scoop or dipper, some 50 percent bigger than any shovel previously built. At that time most people thought the 3/4- or 1-yard shovels seen in the streets were big machines. To construction men, a 3-yard machine was big and heavy. To surface miners, a 12-yard rig was a fair-size machine in those days. But a shovel with a 65-yard dipper? Unbelievable!

The heaviest machine ever to move on land was the Marion 6360 *Captain* stripping shovel, which worked at the Captain Mine of Arch Coal Inc. At just under 15,000 tons, the behemoth carried a dipper of 180 cubic yards. Note the wheel loader parked between the crawlers. The 16-foot clearance allowed coal haulers to pass under the machine.
—Keith Haddock

Of course we now know that the machine making headlines was none other than the famous *Mountaineer*, or Marion 5760, built by the Marion Power Shovel Company of Marion, Ohio. The world had seen nothing like it before, and the designers were moving into unexplored territory.

Report on the 1955 Cleveland Coal Show, held before the shovel erection commenced:

EVEN COAL MEN AMAZED BY SIZE OF NEW MARION STRIPPING GIANT.

Coal industry executives got a sneak preview of Marion's giant at the Coal Show in Cleveland, Ohio, when they saw parts, pictures and artist's drawings of the 5760 coal stripping shovel. The sheer size of the 5760 somewhat amazed coal men, who are already well acquainted with big machinery. The shovel being built for the Hanna Coal Company is the biggest land vehicle in the world. It even compares favorably in weight to the latest U.S. Navy destroyers. But unlike a destroyer, which has a crew of 250 men, the 5760 is completely controlled by just one man.

Fifty percent larger than any existing equipment of its kind, the 5760 is about 100 times larger than the common power shovel normally seen in neighborhood building projects. Each of its eight crawler assemblies weighs 50 tons. In fact, each crawler shoe tips the scales at 840 pounds.

On January 19, 1956, the *Mountaineer* was officially presented to members of the press, radio, and television. Built as tall as a 16-story building, at a cost of $2.6 million, the world's largest power shovel was ready for work. Digging, swinging, and dumping motions were demonstrated to the press and dignitaries before the machine headed out on a 1-mile walk to the coal pit for final commissioning and adjustments. On January 30, the shovel went into full production. The era of the super strippers had begun.

The Marion 5760 boom was 150 feet long, and its dipper handle 92 feet long. Its lifting power of some 250 tons coupled with its wide working range allowed it to move a dipper load of 65-cubic-yards of material from the working face onto a 100-foot high spoil pile some 300 feet away. This performance is comparable to scooping up a load of 66 3,000-pound automobiles, swinging it a distance almost the length of a football field, setting it down on the top of a 10-story building, and swinging it back for another load—all in 50 seconds. The 65 cubic yard dipper could hold 100 tons of material. That's enough to fill a room measuring 11 feet x 20 feet x 8 feet high. The huge shovel had an operating weight of 2,750 tons, and could move an amount of earth and rock approaching three times this weight of earth and rock each hour.

The *Mountaineer* was purchased by the Hanna Coal Company, now known as Consolidation Coal Company (Consol), for work in the company's Georgetown No. 12 Mine, near the small town of Cadiz, Ohio. Hanna had been surface mining coal in the same area since 1940. The new machine joined a fleet of six other stripping shovels consisting of a 12-yard Marion 300, a 24-yard Bucyrus-Erie 550-B, and four Marion 5561s ranging from 35 to 40 yards capacity. All of these shovels continued to operate when the 5760 started work. The Marion 5561 shovels were the world's previous largest.

Fabrication of the machine was started in Marion's shops in October 1954, after many months of preliminary work by engineers from Hanna, Marion, and General Electric (supplier of the electrical equipment). Erection of the machine began when the first crawler was set in position on June 2, 1955. The shovel was erected by Hanna employees, supervised by Marion representatives. It took 125 railroad cars to transport parts to the erection site.

The *Mountaineer* was powered by 14 large DC electric motors providing independent power for each motion. There were four hoist motors, two crowd motors, four swing motors, and four propel motors. The two main AC driving motors totaled 4,650 horsepower. Total peak power demand of the machine was 6,840 kilowatts, enough electrical energy to power 5,200 average homes. The machine received its power via a trailing cable carrying 7,200 volts, the highest voltage yet supplied to a mining machine. As in all large stripping shovels, the *Mountaineer* was controlled by one operator using just two hand levers and two foot pedals. The right lever controls hoist (up or down), the left lever controls the crowd (in or out), and the two pedals control swing (left or right).

Artist's rendering of the Marion 5760, the machine that launched the era of the giant strippers. Shown here with Paul Bunyan, the *Mountaineer* was probably the most famous of the big stripping shovels. It broke all records with its 2,750-ton operating weight when it went to work in 1955.

—Dimitrie Toth Jr. collection

The 65-yard *Mountaineer* at work for the Hanna Coal Company (now Consol) in the southeastern Ohio coal fields. The Marion 5760 featured a passenger elevator inside its 6-foot-diameter center pintle.
—Eric C. Orlemann collection

The machine could be operated from either of two control cabs at the front corners of the machine, three stories above the ground. The idea of the two cabs was to allow the operator to choose the cab on the side of the machine where he was dumping. Telephone communication points were located at strategic points throughout the machine, and radio communication kept the operator in touch with other mobile equipment and supervisors on the site.

The two hoist cables, which pull the dipper upward through the earth and rock overburden, were 2-1/2-inch-diameter steel ropes, each 580 feet long. The herringbone driving gears operating the twin hoist cable drum were 13 feet in diameter. The basic design of the *Mountaineer* retained most of the proven stripping shovel features developed by Marion over the years, including the eight-crawler, hydraulically leveled lower frame, knee-action front end, and hydraulic steering. Each of the eight track sections was 23 feet long and 7 feet high.

It's amazing to find that computers assisted the design engineers on the 5760, a machine designed so long ago—but that was certainly the case. Such a powerful machine demanded cutting-edge computer technology to process hundreds of formulae to select the electrical power system for the operating requirements. This job was accomplished in only two weeks, whereas manual calculating would have required an estimated three years.

Of all the amazing facts told about the *Mountaineer*, the single feature that most captivated the public was the three-passenger elevator installed in the 6-foot-diameter center pin, the first time one had been installed in a shovel. That was actually an excellent safety feature. For the first time, personnel could board the machine without climbing a stairway and having to step from the stationary lower deck to the revolving upper deck while the machine was in motion. The elevator became a standard feature on all subsequent stripping shovels.

Hanna Coal Company determined its need for a shovel the *Mountaineer*'s size through a detailed analysis of the company's coal reserves and stripping requirements. Hanna was known for pioneering new methods, and maximizing the recovery of its reserves at the lowest cost to customers. The company's open pit mining operations involved contour stripping, which entails uncovering the coal seam with one pass of the shovel to the maximum depth of the shovel's operating range. Prior to the advent of the *Mountaineer*, the largest existing shovels, the Marion 5561s, could handle overburden up to about 90 feet. After the final cut had been made by the 5561 shovels, all of the uncovered coal was loaded out and the shovels moved on to new areas within their capability.

In the extensive area stripped in this manner, many miles of "final" high walls were left with the virgin seam exposed. A shortsighted management might have seen this as an ideal chance to recover 150 to 200 feet width of coal from under the high wall by the relatively cheap method of augering where perhaps only 50 percent of the coal would be recovered. But the conservation-minded management of Hanna Coal left the old pits open, to be mined in the future when technology would allow efficient mining of the entire coal seam by larger equipment.

As far back as 1945, company officials began studying the development of a shovel with greater range and capacity than the existing 5561s. Such a unit would be used to reopen the abandoned 5561 cuts and extend the overburden stripping depths to a new maximum of about 120 feet. The company considered it essential that the stripping costs per ton of coal produced not exceed the stripping costs for the smaller shovels, or of deep mining the same coal using the most modern underground equipment available.

The Marion 5760 *Mountaineer* had dug its way out of marketable coal by 1983, and awaited it fate in its final resting place on reclaimed land just south of I-70 near St. Clairsville, Ohio. A few years later the famous *Mountaineer* finally met the scrapper's torch.

—Keith Haddock

Big Hog, the first Bucyrus-Erie 3850-B, removing overburden at Peabody Coal Company's Sinclair Mine, Kentucky. When launched in 1962, the 3850-B was the largest shovel of any type ever built. Its dipper had a capacity of 115 cubic yards.

—Eric C. Orlemann

Many studies were made over relatively long periods of time. Finally in February 1952, the Hanna people decided that the new shovel should have a 150-foot boom, and a dipper size somewhere between 50 and 60 cubic yards. The requirements were handed over to Marion, whose engineers began the detailed design. During the meticulous design process, Hanna and Marion engineers held many on-site meetings to ensure that the designers were totally familiar with similar shovels in actual service. This close contact with the field was very important since the nature of the task involved pushing technology to the extreme. After all, no other mobile machine of such proportions had ever been built.

The detailed and meticulous planning paid off, because after the *Mountaineer* went to work, very few problems were encountered with its operation, and only a few minor modifications were necessary. After the initial break-in period, during which the machine entered and widened the previous cuts, it went on to break all stripping records. During October 1956, the *Mountaineer* moved 2,074,376 cubic yards, a world record at that time for any shovel. In the first 11 months of operation, the machine dug 17,062,094 cubic yards, and set another record. Part of the increase in production was attributed to the *Mountaineer*'s super power, allowing it to cycle in less than 1 minute, compared with 1 minute and 10 seconds, or 1 minute and 30 seconds for the older shovels.

Two different dippers were utilized on the *Mountaineer*, one flat fronted with the dipper teeth in line, the other with a rounded front and the teeth positioned on a curve. Hanna engineers had carried out earlier tests, and found that a rounded dipper not only filled better, but consumed less power in hard digging. Thus the preferred rounded dipper gave higher productivity. Although the initial specifications and magazine articles state the *Mountaineer* dipper held 60 cubic yards, both its dippers were in fact rated at 65 cubic yards when they went into service.

In line with the coal company's farsighted approach to planning, even the new high walls left by the *Mountaineer* were not sterilized in any way. The pits were left to be reopened yet again by even larger equipment not then even envisioned. (The *Silver Spade* and *Gem* super strippers arrived in the mid-1960s to continue mining deeper coal reserves.)

The *Mountaineer* made headlines again in January 1973 when it and one of the 5561 shovels crossed Interstate 70 just west of St. Clairsville, Ohio, to gain access to a new pit. This turned out to be its final working place. After serving its owners well for its entire working career of 23 years, literally moving mountains, the giant shovel was put to rest as a stationary landmark in 1979. She was finally cut up for scrap in 1988.

The Bucyrus-Erie 3850-B, the *Big Hog*

The *Mountaineer*, first of the super strippers, had been working just over four years when Bucyrus-Erie Company announced it had been awarded a contract to build the world's largest shovel. It issued the following press release: *(see page 42)*

The *River King* 3850-B carves its way through rich coal fields in southern Illinois. This second 3850-B went to work in 1964 and had a 140-cubic yard dipper.

—Eric C. Orlemann collection

Left: A crawler assembly for the 3850-B is inspected in Bucyrus-Erie's South Milwaukee, Wisconsin, shops. Each crawler pad weighed 3,700 pounds, and it took 296 pads to shoe the machine's eight crawler assemblies.

—Bucyrus International, Inc.

Bucyrus-Erie to Build Record Shovel

A contract to build a mammoth stripping shovel—largest mobile land machine ever built—has been awarded to Bucyrus-Erie Company, according to an announcement today by President Robert G. Allen. The shovel will be built for Peabody Coal Company, St. Louis, Missouri, for a new mine in western Kentucky. It will be more than twice the size of any shovel now in operation. Allen called the new machine, "a major breakthrough in stripping shovel design." He said the Peabody shovel will have a dipper capacity of 115 cubic yards.

"Merle C. Kelce, Peabody president, stated the purchase of this shovel is a continuing part of Peabody's modernization program. It is expected to lower the cost of mining deeply buried coal, which otherwise could not be recovered by the strip mining method.

"Power requirements for this electrically operated machine are equal to that required for a city of 12,000 people. Fifty-two electric motors, ranging from 1/4 horsepower to 3,000 horsepower each, will operate and propel this giant stripping shovel. The machine will be controlled by a single operator located in his air-conditioned cab five stories up. A passenger elevator will provide access to the cab.

"The Bucyrus-Erie engineering team assigned to the 115-yard shovel offered the following statistics to underscore the great size of the machine:
- In 50 seconds, the shovel will be able to pick up 173 tons of material, dump it 420 feet (more than the length of a football field) away and swing back for the next bite.
- Overburden moved in one month could fill all the cars in a train stretching from Pittsburgh to Chicago.
- The shovel boom will tower 210 feet—as high as the deck of the Golden Gate Bridge, 56 feet higher that the Statue of Liberty, and 45 feet higher than Niagara Falls."

Looking up the boom of the 3850-B. Note the sheaves for the rope crowd system and the massive saddle block, through which the dipper stick moves. The Bucyrus-Erie crowd system features a tubular dipper stick and a two-part boom pinned at its center. The pins are visible at the bottom of the picture.

—Keith Haddock

Following the press release, Bucyrus-Erie's sales department launched the most extensive publicity campaign in the company's history. A potential readership of 50 million people was reached by a "Project 3850-B News Kit," which was mailed to more than 100 major daily newspapers and 500 national trade magazines. The campaign even extended to President Eisenhower, who was presented with a 1/200-scale wood and aluminum working model.

Like all coal companies in the early 1960s, Peabody Coal's shallow reserves were running out and the company was looking at ways to uncover coal under greater depths of overburden without increasing its cost per ton. Larger equipment was the obvious answer, and this led the supershovel contract to Bucyrus-Erie.

The 3850-B, christened *Big Hog*, went to work in August 1962, at the Sinclair Mine in western Kentucky. Newly opened, the mine would supply coal to the first steam generating units at the new Paradise Generating Station, located next to the mine. The Paradise plant's large appetite for coal, the mine's geology, and the type of material to be excavated dictated the size and capability of the shovel. Designing a suitable machine required a high degree of cooperation between the manufacturer and the mine's planning engineers. Peabody Coal's engineers, with their long experience in strip mining, made significant contributions to Bucyrus-Erie's design work on the 3850-B.

Engineers from the two companies specified dimensions for the new shovel: A dumping height of 150 feet; an operating radius of 210 feet; and a dipper capacity of 115 cubic yards. It would be able to remove overburden from a face as high as 100 feet, and move 3 million cubic yards of overburden per month, or 36 million cubic yards every year. This would uncover 14,000 tons of coal every working day.

Another design consideration was the ground pressure, as a stuck machine the size of the 3850-B was unthinkable. So in the first place, the machine was always operated on the firm coal seam, and secondly, the ground pressure was kept below 50 psi. To accomplish this, a standard eight-crawler lower frame was designed, with each crawler 40 feet long and 8 feet high. Each crawler frame carried 37 shoes, for a total of 296 on the machine. Each shoe was 7 feet 6 inches wide, and weighed almost 2 tons. The overall dimensions of the lower frame assembly were 88 feet long by 70 feet 6 inches wide.

Each of the eight crawlers was equipped with its own 250-horsepower propelling motor. The resulting

Shown here under erection at the River King Mine, Illinois, is the second 3850-B belonging to Peabody Coal Company, which also purchased the first. This action shot shows one of four giant hydraulic leveling cylinders being maneuvered into position at the corner of the lower frame. The derrick crane hoists the cylinder as the dozer pushes the lower end into position. After the cylinder is located in the frame, the crawler assembly to the left will be walked in to support it.

—Bucyrus International Inc.

2,000 horsepower could provide the machine with a top speed of only 1/4 mile per hour—fast enough for the world's largest land machine. The 3850-B did have speed where it counts, though. During the swinging part of the digging cycle, its dipper reached a speed of 25 miles per hour. In addition to the eight propel motors, 16 other main DC motors provided independent power to the 3850-B's digging motions. There were 8 for hoisting, 6 for swinging, and 2 for crowding. These main drive motors totaled 11,000 horsepower.

The vertical leveling cylinders at each corner of the machine were 54 inches in diameter. The crawler assemblies were steered using 19-inch-diameter hydraulic cylinders, each capable of exerting a 400-ton force. The hydraulic system contained 4,500 gallons of fluid. The roller bearing, or swing circle, on which the rotating frame swings on the lower frame was 54 feet in diameter. It contained 80 rollers, each 22 inches in diameter.

The massive dimensions of the second 3850-B's 140-yard dipper are dramatically shown in this picture taken shortly after the machine went to work in 1964.
—Bucyrus International Inc.

The engineers studied the basic design of the overall machine on a computer that analyzed all variations of loading during digging and propelling. Including approximately 1,500 tons of ballast added to the rear of the revolving frame, the working weight of the entire machine was estimated to be 9,000 tons.

Bucyrus-Erie manufactured the 3850-B at its South Milwaukee, Wisconsin, plant. It took 300 railcars to transport all components to the Kentucky site, the largest pieces weighing 50 tons. At the erection site, the parts were lifted into place and carefully welded or bolted together so that the machine grew from the ground up. The main lifting equipment at the site was a stiff-leg derrick supplemented by several crawler and truck cranes. The erection took place in a below-ground pit, so overall crane lifting height could be minimized.

Bucyrus-Erie Company received the 3850-B shovel order from Peabody Coal Company in February 1960. Then followed a 30-month period of intense activity to design, fabricate, transport, and erect the greatest machine the world had ever seen. Actual erection on site took 11 months, and the 3850-B *Big Hog* finally dug its first dipperful in August 1962.

During operation, the 3850-B performed exactly as planned, and only a few design modifications were necessary. One example was the 180-foot-long tubular dipper handle, which had to be strengthened. Even though this massive component was made of 3-inch-thick rolled steel, it still started to buckle with its constant travel over the shipper shaft rollers. An additional 6-inch plate was welded along the bottom of the 7-foot-diameter tube.

A more serious event happened to the 3850-B on March 4, 1980. Just as the night shift was ending at 7:15 A.M., the boom collapsed and came crashing back against the body of the machine. No one was injured, but it took three months, with 135 people working around the clock, to repair the damage.

BUCYRUS-ERIE 3850-B "LOT 2"

A little over a year after the first 3850-B was announced, and before the machine went into operation, Bucyrus-Erie received an order for a second

Close-up view showing the machinery house, operator's cab, and eight-crawler undercarriage of the River King 3850-B. The circular openings at the front of the house are the air filtration system intakes. —Eric C. Orlemann

Big Hog dug its own grave! After digging the last cut at the Sinclair Mine, the 3850-B was not cut up for scrap, but buried. In this picture taken in 1986, the shovel is buried to just above roof level. A dragline and two bulldozers are breaking off the gantry legs, which have already been cut—a most undignified way to end the life of a former world record-beater.

—Keith Haddock collection.

The Bucyrus-Erie 3850-B *Big Hog* Starts Work
Extract of the press release of August 29, 1962:

PARADISE, Kentucky—A huge stripping shovel, the largest mobile land machine ever built, began operation here this week, signaling a significant advance by the resurging coal industry. The giant machine bites off roomsful of rock and dirt with its 115-cubic-yard dipper to uncover seams of coal at the Sinclair Mine of Peabody Coal Company, St. Louis. It was designed and built by Bucyrus-Erie Company of South Milwaukee, Wisconsin.

The sheer size of the new shovel staggers the viewer and underscores the very reason it was built. Peabody's coal reserves in this western Kentucky area were deeply buried. Until this machine was conceived, it was not practical to remove all the material above the coal.

The new shovel dwarfs any similar piece of equipment in existence. It towers as high as a 20-story building, removes more than 100,000 cubic yards of material and uncovers more than 14,000 tons of coal each day. In a single month, it will move enough material to fill a train stretching from Pittsburgh to Chicago.

Perhaps the most surprising fact about the record-size shovel is the simplicity of its operation. One man, sitting in his air-conditioned control cab five stories up, controls all movements with just two hand levers and two foot pedals. With a gentle touch he can put his huge machine to work moving 173 tons of dirt and rock in the dipper at 25 miles per hour. An elevator carries the operator to his work. Peabody will operate the machine around the clock, using three operators on eight-hour shifts.

Designing and engineering the spectacular shovel required so many mathematical computations that Bucyrus-Erie put a computer to work on the job. Manufacturing, too, posed its own peculiar problems because of the size of the various components. The cutting edge of the giant dipper, for example, required the largest foundry pour in the company's history.

The original 180-yard dipper of the *Captain* shovel breaks through the bank close to the mine supervisor. The dipper was equipped with two doors, each weighing 15 tons.

—Eric C. Orlemann collection

3850-B from the same customer. This time Peabody wanted a new shovel for its River King Mine in Illinois, with still greater productivity than expected from the first 3850-B, then being built at Sinclair. With a boom some 10 feet shorter than the first machine (200 feet), the dipper capacity was increased to 140 cubic yards. This meant that the title for the world's largest shovel was held only briefly by the Sinclair machine, then passed on to the River King machine. The fact that Peabody signed another contract for a second supershovel, before the first one had even been proven, shows the high degree of confidence the company had in Bucyrus-Erie.

The River King 3850-B was designed to move 44 million cubic yards of overburden annually, uncovering 5 million tons of coal. It followed similar design principles to those employed on the first 3850-B, and incorporated familiar Bucyrus-Erie features established on earlier shovels. These included a tubular dipper handle, crowd motion operated by cables from machinery mounted on the deck, and a two-part boom pinned at its midpoint and tied back to the gantry. The Lot 2 machine, with its 200-foot boom, had a slightly smaller working range than the earlier machine, but weighed a little more, at 9,350 tons including ballast.

Like the first 3850-B, shipment of the parts from the manufacturer's plant in South Milwaukee, Wisconsin, took 300 railcars, and erection at the site took 11 months. The shovel started to dig on August 13, 1964. An industry magazine described the shovel's size in the following report:

> The sheer size of the latest 3850-B shovel staggers the viewer. Described as a battleship in a corn field by local residents, the machine is fast becoming one of the biggest camera attractions in the St. Louis area. The huge machine has a weight equivalent to 6,000 average family automobiles, and is wider than an eight-lane highway. It has a dipper whose cavernous volume is big enough to hold eight Ford Falcons.
>
> Power for the shovel's 100 electric motors, ranging from 1/4 to 3,000 horsepower, is supplied through a 7,200-volt trail cable. During 24 hours of operation, the machine uses as much electricity as a town of 15,000 people.

After a long and productive life at their respective locations, and despite efforts to preserve the mighty shovels as coal interpretive centers, both 3850-Bs were scrapped in a most undignified manner. *Big Hog*, the Sinclair machine, finished work in 1985 after 23 years of digging. The following year, after removal of all toxic and hazardous materials, the once-proud, once-largest shovel, which had provided so much cheap energy to U.S. citizens, was unceremoniously buried in the last pit it dug.

The Lot 2 River King machine fared no better. It finished work in September 1992 after moving over 731 million cubic yards of overburden, and was scrapped the following year. This machine actually had the same day shift operator, Jim Pagliai, for its entire working life. He took control of the machine on its very first day of operation, and was on the machine during its last operating days in 1992, 28 years later. The River King 3850-B turned out to be the most productive and heaviest shovel ever built by Bucyrus-Erie Company.

MARION 6360, THE *CAPTAIN*

The race for the largest stripping shovel ended in 1965 when Marion broke the final record for shovel size.

The 15-foot-high crawler tracks of the 6360 *Captain* shovel are inspected by a mine supervisor. It took eight of these massive crawlers to move the 15,000-ton weight of the world's heaviest land machine.
—Eric C. Orlemann

The incredible Marion 6360, named the *Captain*, was purchased by the Southwestern Illinois Coal Corporation (now Arch Coal, Inc.) to work at its Captain Mine near Percy, Illinois. With an operating weight estimated at 15,000 tons after additional modifications, this behemoth was truly the captain of all shovels, and the heaviest single machine of any type ever to move under its own power on land.

In its November 13, 1965, issue, *Business Week* printed the following account of the mighty machine:

> It stands 21 stories high, it weighs around 15,000 tons and it rolls over the ground under its own power. It is manned by one engineer and one oiler. No, it's not a tall railroad train—it's the most powerful excavating machine ever built, and it just happens to be also the largest self-propelled "mobile" land vehicle.
>
> Marion Power Shovel Company built the super scoop for Southwestern Illinois Coal Corporation's Captain strip mine near Percy, Illinois. Its bucket takes 180 cubic yards of overburden at each bite; that's 300 tons of earth and rock. And it takes a mouthful every 1 minute and 12 seconds.
>
> The electric-powered shovel dug its first dirt in mid-October close to its assembly site. Mine officials expect it will be active 85 percent of the time, 24 hours a day, seven days a week, uncovering enough coal each day to fill 155 to 165 railroad cars.

When the Captain Mine opened, the company's engineers considered a number of different methods to uncover the property's two seams of coal. It was determined that the most economical operation would be to employ a single large shovel. The big stripper would stand on the lower seam and strip about 90 feet of material from the top seam at the side of the machine. From the same position, the shovel would then remove another 25 feet of overburden or parting from the lower seam by digging straight ahead. In this way, both seams would be uncovered by a single pass of the machine from one end of the cut to the other.

Designed to meet the above pit conditions, as well as the mine's coal production requirements, the 6360 carried a 180-cubic-yard dipper on a 215-foot boom, with a maximum dumping radius of almost 220 feet—specifications exceeding those of any previous machine. During the tendering process for the shovel order, Bucyrus-Erie proposed its similar-sized 4850-B, but this model was never built.

Overall, the 6360 shovel looked like most other Marion stripping shovels, with its eight-crawler hydraulically leveled lower works and knee-action crowd front end. However, what set this shovel apart was its gigantic proportions. Even the dipper door was twinned, each separately hinged half weighing 15 tons! This was the only shovel to have two dipper doors.

During erection, the 6360 presented new problems not encountered in smaller machines. One of

Two record-beaters in the same cut! The Marion 6360 *Captain* shovel is digging the lower overburden and the parting between two seams of coal while the massive 5872-WX cross-pit bucket wheel excavator, largest of its type, removes the upper overburden. The machines are digging from right to left.
—Eric C. Orlemann

The Marion 6360 *Captain* shovel was destroyed by a disastrous fire in September 1991. In this picture, the stricken monster awaits the scrapper's torch in late 1992. During its working life, from 1965 to 1991, the *Captain* shovel moved over 809 million cubic yards of overburden.
—Eric C. Orlemann

the most critical was to design component parts within practical limits for manufacturing, shipping, and erection, while still maintaining structural stability in the field. The first step was to design lower frame and crawler assemblies strong enough to support the immense weight of the machine, while at the same time not exceeding the specified ground bearing pressure of 60 psi under the crawler shoes. These requirements dictated the size and dimensions of the lower frame and crawlers. The lower frame consisted of 29 sections welded together to form a 63-foot square box 14 feet deep. High strength steel plates up to 4 inches thick were used.

At each corner of the huge frame, the designers put jack assemblies to level the machine. These massive hydraulic units also served as giant legs to support the machine on the crawler assemblies. The vertical 66-inch diameter cylinders were welded integral with the frame, and each matching piston was attached with a pivot point at its lower end to a huge axle, 30 feet long, carrying the two crawler assemblies. This axle was so large that Marion had to design a new shop tool to machine it.

To achieve the specified ground pressure, each pair of crawlers was 45 feet long, 30 feet wide, and 16 feet high. Adequate ground support was provided by 10-foot-wide crawler shoes, each weighing 3-1/4 tons. A complete crawler set for the machine required 336 of these shoes. To propel the machine, Marion developed a new drive system with a drive sprocket at each end of the crawler frames. By using two separate gear trains and driving each with a 200-horsepower AC electric motor, the power transmission parts were kept within practical limits for manufacturing and maintenance. Even then the final drive gear in each train was over 5 feet in diameter. The drive sprockets were impressive too. They were 78 inches in diameter and weighed 8-1/2 tons. Steering the 6360 was accomplished by four massive horizontal hydraulic cylinders, one for each of the four crawler assemblies. The *Captain* shovel could travel on its four pairs of crawlers at a top speed of 1/4 mile per hour.

In addition to the above-mentioned AC propel motors totaling 3,200 horsepower, electrical equipment inside the house included eight motors to power the two hoist drums, four motors to power the crowd motion, and eight motors to rotate the machine. These were all DC motors with a combined total rating of 15,000 horsepower.

Field erection of the 6360 *Captain* took 150,000 man hours over an 18-month period. Crews from Bollmeier Construction & Engineering Company, the erection contractor, were supplemented by mine personnel and as many as 15 supervisors from Marion Power Shovel. The *Captain* dedication ceremony was held on October 15, 1965.

The *Silver Spade*, one of two Bucyrus-Erie 1950-B shovels purchased by Hanna Coal Company (now Consol), is shown at work near New Athens, Ohio. Carrying a dipper of 105 cubic yards on a boom 200 feet long, it first dug dirt in 1965 and is still operating intermittently at the time of writing.

—Keith Haddock

The *Captain* shovel joined one of the largest fleets of earthmoving equipment ever assembled in one place. At its peak in the early 1980s, the mine operated over 300 pieces of heavy equipment, including three other stripping shovels (25, 40, and 65 yards in capacity), seven draglines ranging up to 100 yards in capacity, four bucket wheel excavators with associated conveyors and spreaders, and five coal loading shovels. Mobile equipment included 35 bulldozers, 23 haul trucks, 12 wheel loaders, 13 scrapers, and 10 drills.

After start-up and some initial mechanical problems, the *Captain* went on to lead a successful career. But maintenance crews had to learn how to tackle the many unique problems arising in a machine of such gargantuan proportions. A crane or other heavy lifting equipment was needed for even a "small" job; virtually nothing could be manhandled, from the smallest pin to the boom point sheaves. Some major components, such as the crawler assemblies, were changed out during the life of the machine. With a spare crawler unit on hand, the old unit was simply walked out under its own power, and the replacement unit walked back under the machine.

During operation, the *Captain* was assisted by a wheel loader assigned to clear stray rocks away from the crawler tracks. When not in use, the loader was parked under the shovel. The 16-foot clearance height under the shovel allowed a haul road to pass between its tracks, so the coal haulers drove right under the shovel when hauling from the far end of the pit. As they passed through, they slowed or waited as the massive dipper carrying 250 tons of earth and rock passed overhead toward the spoil pile, and then returned empty with its two doors swinging. By allowing the coal haulers free access under the shovel, the company saved the cost of building an extra ramp at every pit end.

In the pit, the two-seam, single-pass method worked well, but as the overburden increased in height by the late 1970s, additional equipment was brought in to reduce the height ahead of the *Captain*. Initially dozers worked on top of the face to push material down to the shovel, but later the mine employed an expensive system consisting of two O&K bucket wheel excavators, each connected to a Weserhutte round-the-pit conveyor belt which fed into a Mitsubishi spreader on the spoil side (one set for each half of the pit). Finally, in February 1986, a new cross-pit bucket wheel excavator replaced the former wheel excavators and conveyors. This was the Bucyrus-Erie 5872-WX, the world's largest cross-pit bucket wheel excavator. Its job was to remove the upper layers of overburden and convey the material directly across the pit onto the spoil pile. The *Captain* removed the remaining overburden down to the coal. Thus two of the world's largest pieces of equipment worked together in the same pit.

The active life of the *Captain* came to an abrupt end on September 9, 1991, when a disastrous fire burned for several hours in its lower works. Thought to have started when a hydraulic line burst and sprayed fluid over live electrical panels, the fire was fueled by a build-up of grease and oil in the swing circle area, and proved very difficult to extinguish.

The fire was discovered by the machine's operating crew, who immediately tried to fight the fire with dry chemical fire extinguishers. Meanwhile, the mine office was alerted and dispatched the mine's two fire trucks. When their efforts failed, a call for help to the Pinckneyville Rural Fire Department at 9:44 P.M. triggered one of the most extensive firefighting activities the area had ever seen. By the time the night was over, fire departments from Pinckneyville, Cutler, Percy, Campbell Hill, Steeleville, Coulterville, DuQuoin, and Sparta were battling the inferno. Observers described the fire as, "looking like a very big gas stove burner with the largest pot you have ever seen sitting on it—the fire was coming out all around the lower frame, and was 30 feet off the ground."

Eight pumper trucks were on the scene, plus three tankers, one aerial ladder, two rescue trucks, one lighting unit, one deluge gun, and 97 firefighters. In addition, numerous pieces of equipment and personnel from the mine assisted with the disaster.

Mine crews fighting the fire with hand lines on the machine above the fire had to be removed in a manlift due to the smoke and heat. Another man was not so lucky. Trapped high on the machine, beyond the range of the available aerial ladders, he could not descend on the shovel's stairways or in the elevator. He

A striking view under the boom of the 1950-B *Silver Spade*. The 105-yard dipper pauses momentarily as the mine supervisor shows off the immense proportions of this super stripper.

—Bucyrus International Inc.

was finally talked into putting on a safety belt and snapping it on to one of the 2-inch boom suspension cables running from the gantry to the top of the boom. He then walked across on two other cables to the boom. Once there, he climbed down to where the aerial ladder could reach him and lower him to safety. The fire was basically out by 2:30 A.M., although smaller fires continued to erupt throughout the night.

After assessing the damage, mine officials considered the cost of repairs too great, and the machine was scrapped in late 1992. During its lifetime, the *Captain* moved about 809,300,000 cubic yards of overburden, over three times the quantity of material excavated for the Panama Canal! A Marion 5900

The 1950-B *Silver Spade* swings above two mine workers. Note the cable drum for winding the trailing power cable and the vertical hydraulic cylinder above the crawler assembly for leveling the machine.
—Keith Haddock

stripping shovel with a 105-cubic-yard dipper was moved from another part of the mine to carry on the duties of the stricken 6360. The Captain Mine ceased operations in 1998. Officials cited the Clean Air Act's sulfur limitations, which caused power stations to switch to low-sulfur western coal.

BUCYRUS-ERIE 1950-Bs, THE *SILVER SPADE* AND THE *GEM*

The *Silver Spade* was not the largest shovel built, nor was it the first or the last stripping shovel put to work. It receives special attention here because, at the time of writing, it is the only supershovel with a potential working life still ahead of it. Although the shovel was idled at the end of 1999, at the time of publication, it is waiting for a mining permit. It can move into a new area when market conditions permit. The now-famous *Silver Spade* has received much publicity, including magazine articles, videos, and television documentaries. Although breaking no records, it ranks high with the largest shovels ever built. It started work in November 1965, just a month after the world's largest shovel, the *Captain*, was dedicated.

The *Silver Spade*, or Bucyrus-Erie 1950-B to use its technical designation, was actually the first of two of this model built by Bucyrus. It was purchased by the Hanna Coal Company, a division of Consolidation Coal Company (Consol), for coal stripping in the Georgetown area of Ohio. The *Silver Spade*, named to commemorate the 25th anniversary of Hanna Coal, joined the *Mountaineer*, which had started work in the same area a decade earlier.

The *Silver Spade* is equipped with a 200-foot boom and a dipper holding 105 cubic yards. It is 59 feet wide at ground level, and the top of the boom reaches 191 feet into the air. Its digging range is 390 feet, meaning that if the shovel stood in the center of a football field, it could pick up a load 45 feet beyond one goal line, swing it around 180 degrees, and deposit it 45 feet beyond the other goal line. The swing circle bearing is 50 feet in diameter and contains 100 hardened steel rollers 16 inches in diameter. The working weight of the machine is 7,000 tons.

The 1950-B is served by a 7,200-volt trailing cable weighing 20 pounds per foot. On board, AC main driving motors totaling 9,000 horsepower drive DC generators for the shovel's main motions—hoist (eight motors), swing (four motors), and crowd (two motors). In addition, there are eight propelling motors, one in each of the eight crawler track frames.

The second Bucyrus-Erie 1950-B was the *Gem*, which stood for Giant Earth Mover. Carrying a 130-yard dipper on a 170-foot boom, it worked in the same mining location as the first 1950-B.

The 1950-B's most distinguishing feature is its knee-action front end, Bucyrus-Erie's only model to be so equipped. The customer insisted that the shovel be equipped with the knee-action front end, so a special arrangement had to be made with Marion, whose patents covered this design. The two companies reached an agreement: Bucyrus could use the knee-action crowd, while Marion could use the Bucyrus-designed cable crowd system instead of its own rack-and-pinion design.

The other Bucyrus-Erie 1950-B shovel was the *Gem of Egypt*. This was also purchased by the Hanna Coal Company for work at the Egypt Valley Mine, not far away from the *Silver Spade*'s location. Its design and operating weight were almost identical to the first 1950-B, except that it carried a shorter boom (170 feet) and a larger dipper (130 cubic yards). It turned out to be the last stripping shovel built by Bucyrus-Erie Company.

The launching of this giant shovel was a very popular event for the people in the area. The public was invited to an "open house" to inspect the machine up close. Bucyrus-Erie's *Scoop* magazine reported on the Gem of Egypt's launch:

Scoop magazine on the Gem of Egypt:

Road signs all along U.S. 40 and other highways announced the open house of the "world's third-largest stripping shovel," and it seemed that every other car turned into the country road that went 3 miles inland to the mine site. Arriving at Egypt Valley Mine, you find that Hanna has fashioned two parking lots for the occasion, both freshly covered with a thick layer of gravel. This was to prove extremely good foresight, for the ensuing rain would have turned the lots into a quagmire. You are marshaled into a registration line, which got pretty lengthy during peak hours the afternoons of Saturday and Sunday. Finally you reach the table where, in return for your signature, you receive literature from both Hanna and Bucyrus-Erie, answering most every foreseeable question.

Now it's into another long line to climb the erected ramp leading inside the *Gem* itself. The line files past the giant dipper, and there's a stir among the people as they look closely at its hugeness. Families pose in it as they would at Mammoth Cave, and cameras click all around. Onward to the machinery house. You can tell people that the house is as big as a three-story, six-family apartment building, but until they roam around it themselves, they cannot feel the sense of magnitude, power, precision, and solidity they experience by seeing and touching. Inside there are floral arrangements from neighboring concerns congratulating Hanna on the open house. People are swarming all over the machine like Lilliputians, poking here and there. All the unsafe places have been roped off, under the direction of the company's safety officer. Before leaving the house, you are met by Hanna staff, who obligingly answer any remaining questions you might have, and if you are a woman or a child, you are eligible for a souvenir balloon, of which Hanna handed out over 10,000.

Outside, a series of circus-like tents housed displays of pictures of mining operations, slide shows of mining reclamation achievements, and most importantly for the visitors—food! Over the weekend, 2,400 pounds of beef, 9,000 buns, and 300 gallons of coffee were consumed. And 10,000 napkins and paper plates were used. In all, over 25,000 people attended the open house, and that despite cold and damp weather.

Left: A spectacular aerial shot of the 1950-B *Gem* in operation. The *Gem* moved 548 million cubic yards of material during its working life, from 1967 to 1988.

—Consol

The front end of the *Gem* shovel utilizes the "grasshopper," or knee-action crowd, the only Bucyrus model to be so equipped. The crowd arm runs between the two halves of the boom, and is pinned to the stiff leg.

—Keith Haddock

The *Gem of Egypt*, second of the 1950-Bs, started work early in 1967. Then in 1974, it moved 14 miles across country to the Mahoning Valley Mine near Cadiz, Ohio, where it remined some areas of a shallower seam previously worked by Consol's smaller shovels. After the move, the shovel became known as The *GEM* (Giant Excavating Machine). The *GEM* was finally parked in 1988 after moving over 548 million cubic yards of overburden. Four years later, the mammoth *GEM* was cut up for scrap.

Dramatic new equipment advances, like the stripping shovels in the 1960s, helped to pioneer cheap mining methods, and boost an ailing coal industry into one supplying the lion's share of electric power generation. Since then coal production has risen steadily into the twenty-first century, and annual coal production today is double that achieved in the 1960s.

PART TWO
WALKING DRAGLINES

CHAPTER FOUR

Walking Dragline Origin & Application

The dragline, like its brother the stripping shovel, is a member of the excavator family of giant digging machines. Draglines are found in sizes ranging from the small convertible (universal) excavator, formerly the backbone of the general contractor's equipment fleet, to some of the largest machines ever to move on land. Universal excavators, now largely superseded by hydraulic excavators, were mounted on crawler tracks or rubber tires, and ranged in bucket capacity from 3/8 yard to about 5 cubic yards. These "construction-sized" machines can be equipped with several different front-end attachments such as backhoe, shovel, clamshell, lift crane, and of course the dragline.

The *Estevan Eagle* is the name of this Marion 8750 dragline at Luscar's Boundary Dam Mine, Estevan, Saskatchewan, where it digs with a 98-yard bucket. A modern dragline designed for the coal boom in the 1970s, the 8750 featured Marion's improved walking cam design with an outboard bearing.

—Keith Haddock

The Bucyrus-Erie 480-W was one of the company's most popular walking draglines. With a production run from 1955 to 1979, 35 of these 17-yard class machines were shipped to many locations around the world. Here, C & K Coal Company's late Model 480-W is stripping coal in the hills around Clarion, Pennsylvania.
—Bucyrus International Inc.

The larger draglines are usually of the walking type, but some crawler draglines have been built up to 21 cubic yards capacity for certain applications. Above this size, all draglines are walkers, and the subject of this book. They range up to the behemoth 220-yard *Big Muskie*, the largest ever built. These mechanical monsters are found in surface mining operations, or in large gravel pits and limestone quarries. Occasionally, contractors at large construction projects, such as canals and dams, have made good use of the walking dragline's appetite for large volumes of earth.

The first walking dragline was built in 1913, and except for machines built in the CIS (former Soviet Union) and China, and one or two prototype machines, the entire world market for walking draglines has been supplied by only four manufacturers:

1. Monighan Machine Company, whose machines became Bucyrus-Erie's line in 1932;
2. Page Engineering, acquired by Harnischfeger Corporation (P&H) in 1988;
3. Ransomes & Rapier Ltd., acquired by Bucyrus in 1988; and
4. Marion Power Shovel Company, acquired by Bucyrus in 1997.

Appendix 4 shows that out of a total of more than 2,000 walking draglines built worldwide, 1,497 have been built in the United States.

When the Marion Power Shovel Company was purchased by Bucyrus International, Inc., in 1997, it reduced the already small number of walking dragline suppliers to just two, plus those still built in the former Soviet Union. Appendix 1 contains a history of all these dragline manufacturers.

Above: The intermediate-sized Bucyrus-Erie 800-W is shown here swinging a 28-yard Page archless bucket. It is working at the Indian Head Mine, North Dakota, for the North American Coal Corporation.
—Keith Haddock

Left: Peabody Coal Company's Hawthorne Mine, in Indiana, boasted two large draglines before it closed in 1999. This is the 95-yard Bucyrus-Erie 2570-W, featuring the cam-and-slide walking system. It worked in tandem with the 155-yard Marion 8900.
—Eric C. Orlemann

Above: A Marion 7500 owned by British contractor G. Wimpey & Company uncovers coal at a site in northern England. This 17-yard electric-powered machine started work in 1971, and worked at three different locations before being scrapped in 1995.

—David R. Wootton

Below: In 1974, Marion shipped the only 7620 dragline to Knife River Coal Company in North Dakota. Named *Queen o' Buttes*, it carried a bucket holding 30 cubic yards on a 235-foot-long boom.

—Keith Haddock collection

This Marion 7820 is presently owned by C & K Coal Company of Clarion, Pennsylvania. It features Marion's simple walking crank system. The 250-foot boom carries a 45-yard bucket.

—Keith Haddock collection

The 1960s was the era of the super stripping machines. That decade's largest stripping shovels and draglines have never been exceeded in size. During the 1970s, when dozens of new walking draglines went to work, mine operators preferred slightly smaller models, built using high technology on proven principles. These draglines proved to be more reliable, and in some cases actually out-produced the much-larger behemoths of the earlier decade.

When the Arab oil embargo hit in the early 1970s, coal production was seen as the savior to the United States' energy crisis. Traditional coal companies expanded their production, but others not formerly in the business purchased coal-producing properties. They all clamored to get their names on the dragline order lists, and those with urgent coal-supply contracts were even willing to pay a premium to advance their delivery date. At its peak, backlogged dragline production stretched out so far that customers had to wait over four years for delivery. Manufacturers rallied to fill the orders, but as soon as they stepped up production and expanded their plants, demand for new machines dropped off drastically, and several customers even canceled their orders. The Middle East panic was over, and the energy situation had returned to normal by the end of the decade.

Like ships on the ocean, and the big shovels described in Part I, the large walking draglines are often given names by which they are referred throughout their working lives (Appendix 2). They have their own character, and are looked upon with affection by their owners and operators.

MACHINE DESCRIPTION

The main four components of the dragline are (1) the circular base or tub, (2) the revolving frame on which is mounted all the draw works, shoes, and propelling machinery, enclosed by the machinery house, (3) the boom and bucket, and (4) the gantry, which supports the boom. A dragline looks like a large crane, but instead of a lifting hook, a digging bucket is suspended from its boom by hoist ropes leading to a winding drum on the revolving frame. Another drum winds the drag ropes, which also connect to the bucket. The revolving frame rotates on a large-diameter bearing made up of multiple rollers that run

The early Page 600 series dragline is represented here by the Model 627, the only one of this particular model manufactured. It carried a 12-yard bucket and worked at the Muskingum Mine in Ohio, the very mine where the world's largest dragline, *Big Muskie*, spent its working life.

—Keith Haddock

between the upper and lower swing rails. A large circular rack is fixed on top of the tub. This is engaged by the pinions driven from the swing motors in the revolving frame to provide the swing motion. Some draglines have as many as 16 swing motors, each with its own gear reduction and pinion.

Dragline booms are suspended in many different ways depending on their length and manufacturer's preference. Smaller walking draglines use multiple-part ropes passing back and forth from the boom point to the gantry. The rope is wound on an independently powered boom hoist drum on the machinery deck. This "live boom" feature is not usually found on the larger draglines. Boom cable suspension usually includes a secondary "safety" rope, which is tied in place at the boom's working angle. This rope is capable of supporting the boom in case of failure of the main rope or the boom hoist mechanism.

On larger draglines, no boom hoist mechanism is provided. The boom on these machines is raised by the main hoist drum. The boom hoist rope is temporarily wrapped around the hoist drum, and reeved through several parts to an intermediate mast that supports the boom. The mast supports the boom by fixed-length "bridge strand" cables, so that the mast and boom are raised in unison. When the boom reaches its predetermined working angle, it is tied back to the "A-frame" or gantry with further "bridge strand" cables, or pinned to a solid connecting structure, depending on the machine design. Once the boom is in place, the hoist rope can be dispensed with, since a dragline does not normally change its boom angle during operation.

Raising and lowering a large dragline boom is a major task, and may occur only two or three

A diesel-powered Page 728 swings a 12-yard bucket for the Snyder Companies near Kittanning, Pennsylvania. It features Page's improved walking system introduced in 1935. The engine is a Page 12-1/2x16 V-type with eight cylinders.
—Keith Haddock

times in a 20-year period. A dragline manufacturer's representative will typically come to the site to oversee the raising and lowering of the boom, since the job is not routine for the mine's maintenance staff. Most dragline booms are equipped with a red light at the top to warn low-flying aircraft to keep clear.

In action, the dragline drags its bucket toward the machine as it collects its load. When full, the bucket is hoisted by the hoist ropes, the machine swings, and the bucket contents are dumped in a pile off to the side. The bucket is arranged to dump automatically when the tension is released on the drag rope. A dragline's boom angle remains fixed during all operations. The boom angle is usually set to provide a certain dumping radius to comply with the original machine specification, which takes into account the boom length and bucket capacity. The mine engineer works with the manufacturer to create a design specification that fits the mine's production requirements, site geology, and geotechnical conditions.

The drag ropes are guided through a fairlead consisting of a series of sheaves that can move to accommodate the constantly changing rope angle between the bucket and the machine as it swings. The fairlead also accommodates the changing fleet angle of the rope as it winds on its drum.

The drag and hoist machinery are termed the "draw works," and these may have four or more DC motors to power each motion. The motors work together, and are geared directly through intermediate shafts to the main "bull" gear on the hoist or drag shaft. No clutches or brakes are needed during digging, as the DC motors allow control of speed and direction through the operator's levers in the cab. The operator may hold hoist, drag, and swing motions

This Page 736 shipped in 1984 was one of the last Page draglines produced before the company was taken over by Harnischfeger Corporation (P&H). It works at the Sheerness Mine of Luscar Ltd. near Hanna, Alberta, where it uncovers coal with a 30-yard bucket.
—Keith Haddock

with air-operated brakes. These brakes are applied at rest, not when components are moving.

Some intermediate-size draglines (10 to 20 cubic yards) use only one DC electric motor to power both the hoist and drag winding drums. In this arrangement, the drums are connected to their driveshafts through air-operated clutches. When released, rotation of each drum is controlled by its own air-operated band brake. The hoist and drag cables are wound on their drums so that when both drums are engaged to turn, one winds its rope in, while the other unwinds its rope. This system is called "synchronous control." Its big advantage, apart from the lower cost of utilizing only one motor, is that when both drums are engaged, the force of the loaded bucket pulling on the drag rope when it pays out helps to hoist the bucket to its dumping position. Similarly, the weight of the empty bucket when being lowered back into the cut helps to wind in the drag rope. This arrangement also results in much reduced brake wear.

The walking system on a dragline is very simple. To take one step, the shoes are rotated in the direction of travel by the walk shaft and an eccentric drive, so that they touch the ground simultaneously. Further rotation of the walk shaft lifts the leading edge of the dragline's circular base or tub off the ground. In this raised position, the machine is supported on three points: the two shoes and the tub trailing edge. Continued eccentric rotation slides the machine ahead a distance of one step (about 6 feet), and gently lowers the machine back on its base. The shoes continue to rotate, and the process is repeated for the next step. Changing direction (steering the

The largest diesel-powered walking dragline was this 26-yard Page 738. One of only two built, it utilized two Page diesel engines—an 8-cylinder powering the electric swing generator drive, and a 16-cylinder driving the main drums through friction clutches. The hoist and drag drum brake bands were 9 feet in diameter.
—Keith Haddock

The Page 762 was the second-largest dragline Page built. The one and only machine was originally shipped to Kaiser Resources at Sparwood, British Columbia, in 1969. The picture shows the 54-yard machine hard at work after relocation to Consol's Glen Harold Mine in North Dakota.
—Keith Haddock

A Rapier W300 working at London Brick Company's Calvert operations in central England in 1989. Featuring Rapier's unique cantilever boom design, the machine carries a bucket of 7 cubic yards, and is electrically operated.

—David R. Wootton

machine) is just a matter of swinging the machine to point in the desired direction when the shoes are off the ground. On the smaller draglines, the shoes are powered by a single large shaft running across the width of the machine, and driven by gearing from the propel motor. Larger draglines use separate walk shafts, one on each side of the machine. An electronic timing device synchronizes the two shoes so they rotate in unison. Walking draglines always walk backward, as it is necessary to walk away from the hole or pit being excavated.

An important advantage of the walking dragline is the very low ground pressure exerted by the large-diameter base on which the machine sits while digging. When walking, only about 80 percent of the machine's weight is transferred to the shoes, which can be made with large dimensions to reduce ground pressure. The other 20 percent of the weight is carried by the tub as it drags along the ground.

WALKING SYSTEMS

Each walking dragline manufacturer has patented its own walking systems, and some have several different types in use on different models. The type selected usually depends on the size of the machine. All systems incorporate a type of eccentric drive to the shoes, which are driven by a walk shaft. Bucyrus draglines up to the 70-yard class use the "Monighan" system invented by Oscar Martinson in 1925 (Figure 4.1). This utilizes an eccentric (cam) wheel running inside an oval track in a frame pivoted to the shoes. As the walk shaft turns slowly, the shoes follow an oval path, which tilts the machine, and gently moves it backward a distance of one step.

Bucyrus draglines larger than 70 cubic yards employ the "cam and slide" design (Figure 4.2). This method also employs an eccentric wheel, but the wheel runs in a circular roller bearing in the walking leg. A tie rod connects the shoe with the circumference of the cam wheel, so that when the latter turns, the cam raises the machine off the ground, and at the same time slides the machine horizontally on a greased rail fixed to the top of the shoe. The world's largest dragline, the 220-yard Bucyrus-Erie 4250-W known as *Big Muskie*, employs a fully hydraulic walk system. It is described in chapter 6.

The Ransomes & Rapier draglines are fitted with the patented "Cameron and Heath" walking system (Figure 4.3). In this design, an eccentric wheel runs in a large roller bearing fixed to the walking leg. As the walking shaft turns, the necessary oval motion is obtained from a tie rod connecting the shoe to a point on the dragline's frame.

Marion uses a simple crank on each end of the walking shaft on its smaller draglines (Figure 4.4). The crank bears into a trunnion that is pivoted to the walking shoe. The other end of the trunnion is attached to the dragline's frame above the walking shaft by means of a tie rod. For machines over 30 cubic yards, Marion uses an eccentric wheel running in a large roller bearing, similar to the Rapier system (Figure 4.5). The Marion 8800 and 8900 models (85 to 145 cubic yards) employ two walk shafts with eccentric wheels on their ends (Figure 4.6). These are connected by heavy links to the same point on the shoe. Since the two

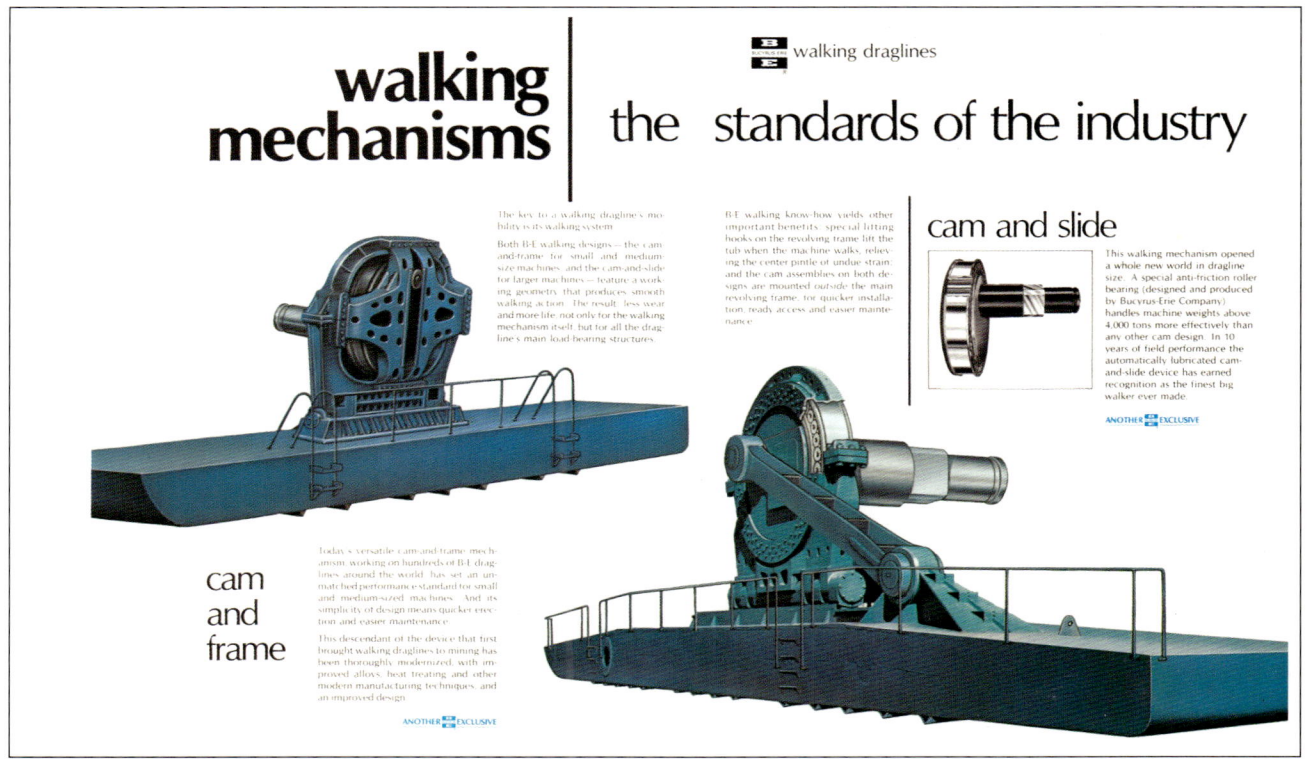

Figure 4.2

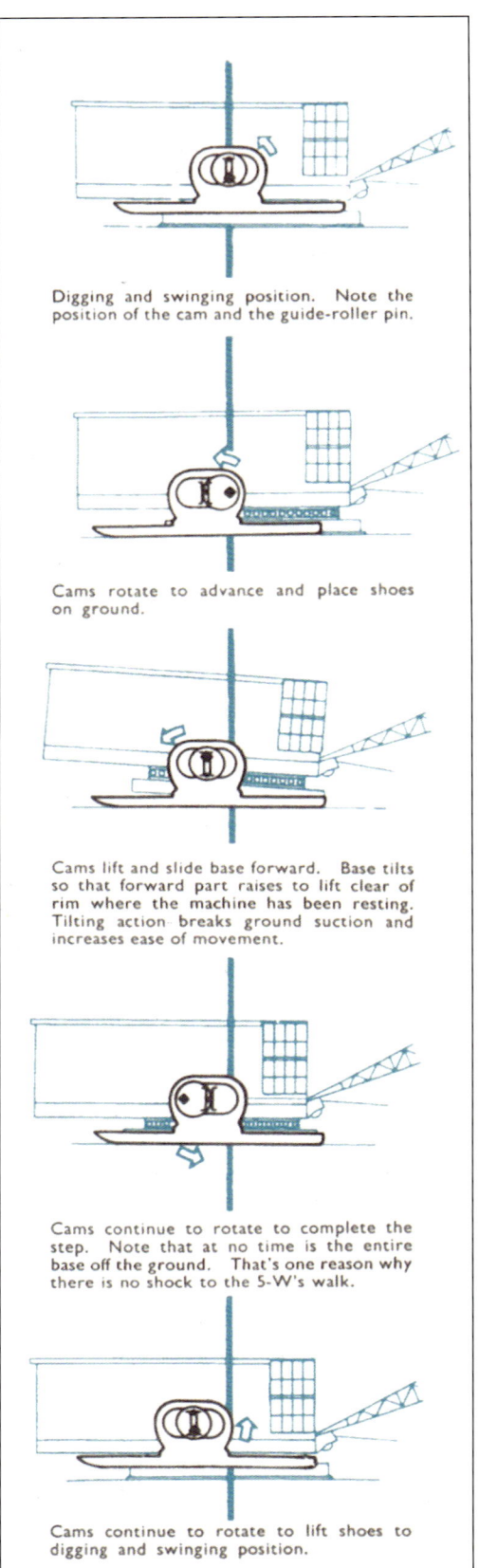

Figure 4.1

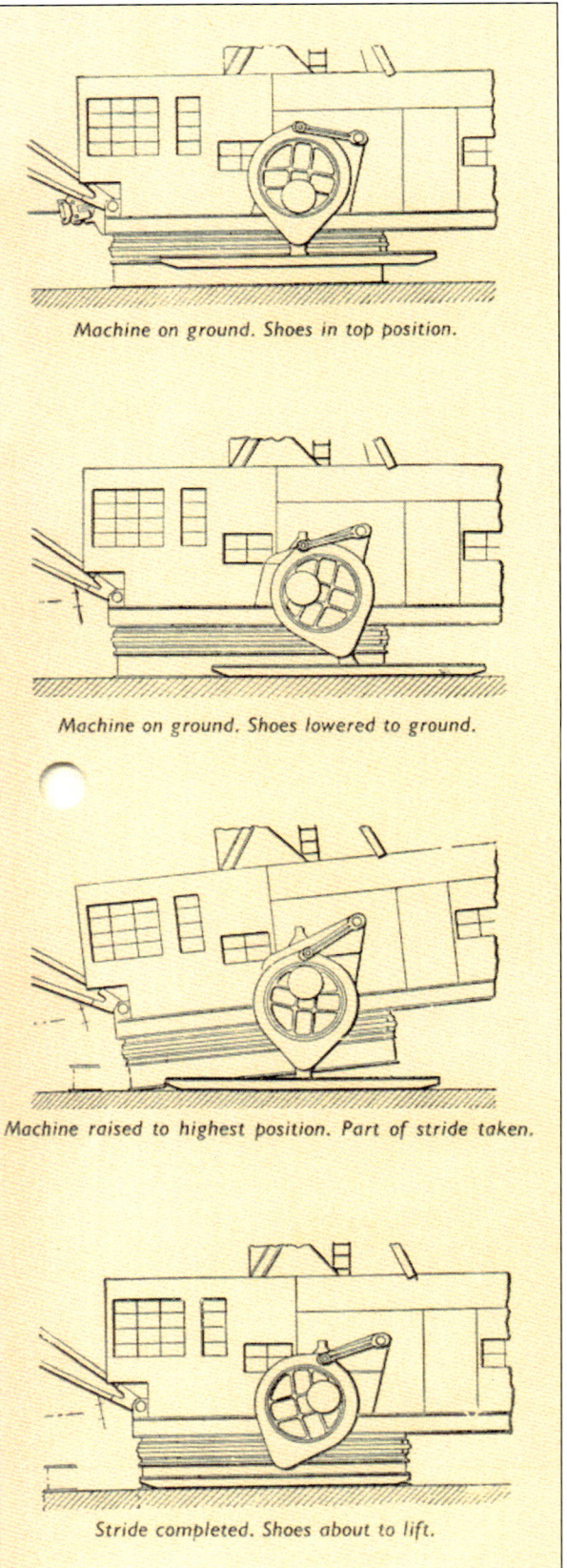

Figure 4.3

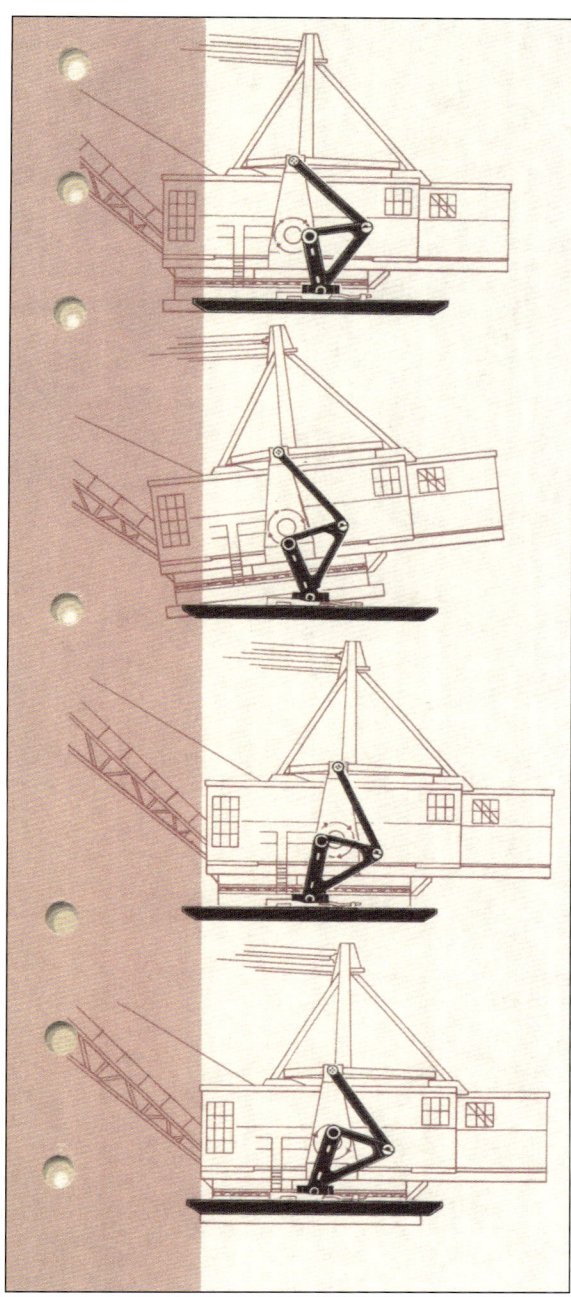

Figure 4.4

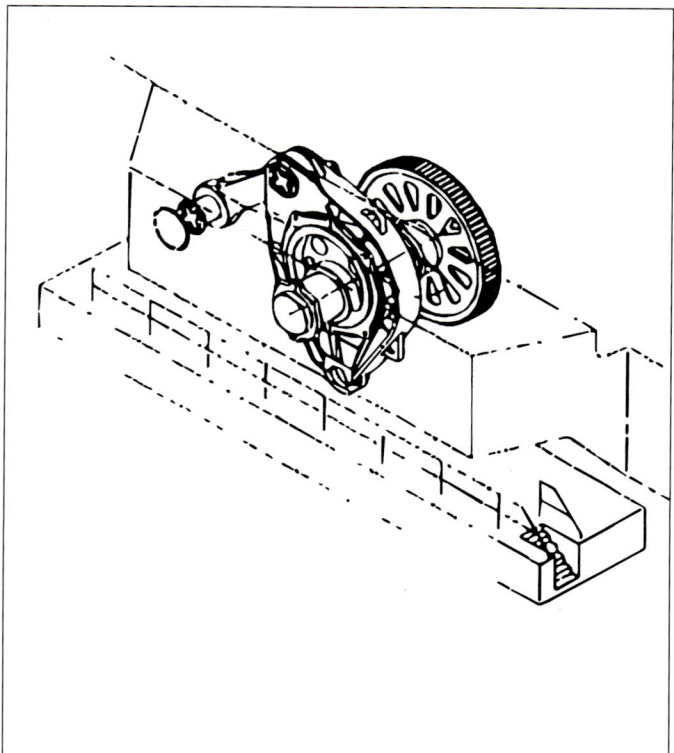

Figure 4.5

eccentrics have different throws, the necessary elliptical motion is imparted to the shoes as the shafts rotate.

All the walking draglines produced by Page Engineering Company since 1935 used a walking spud on each side of the machine to carry the shoes (Figure 4.7). The spud is loosely connected to a crank on the end of the walk shaft via a slotted bearing. The crank also connects to a hanger that carries the weight of the machine during the walking step. The top end of the spud is guided by the upper spud roller, which runs in a slot. As the crank rotates, the shoes move in an elliptical path, necessary for a smooth walking motion.

The new P&H walking draglines being offered by Harnischfeger Corporation feature a cam and roller bearing similar to the Marion and Rapier types. Harnischfeger changed the former Page walking system to this design immediately following its purchase of Page's manufacturing rights.

POWER FOR THE DRAGLINES

From the very first in 1913, walking draglines have been optionally powered by steam engines, diesel/gasoline engines, or electric motors. Steam was chosen by purchasers of some early machines. Diesel power has generally been restricted to machines of under 20 cubic yards capacity, but one diesel machine, a Page 738 built in 1963, swung a bucket carrying a record 26 cubic yards. Electricity did not come into popular use in draglines until the late 1920s, but as draglines grew bigger, all were powered by independent DC electric motors for hoist, drag, and swing motions.

In 1924, Page developed the first diesel engine exclusively designed for dragline use. After that, Page installed its own engines in most of its diesel-powered draglines. They were of the four-cycle, horizontal "V" type, or

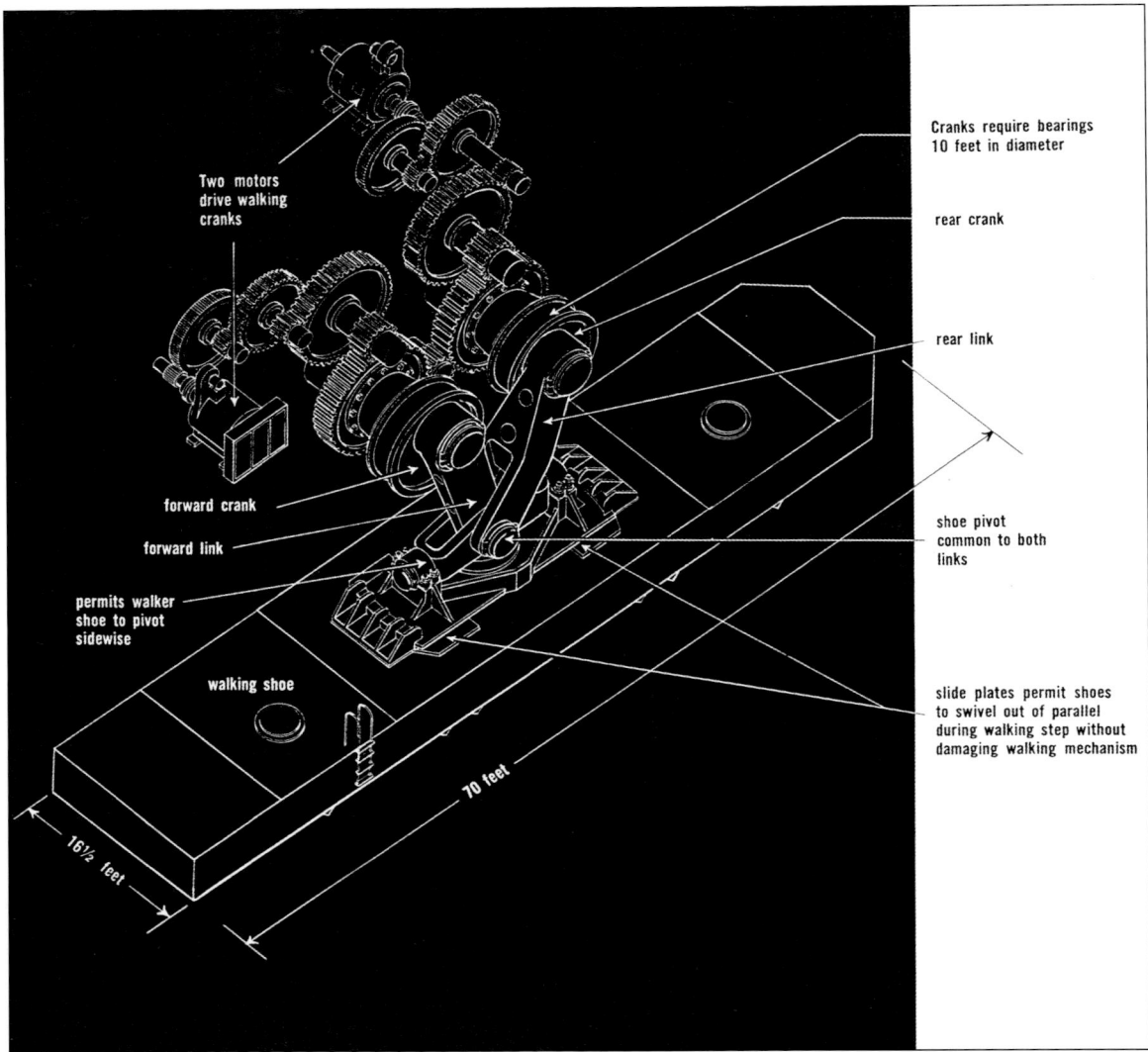

Figure 4.6

horizontal "in-line" type, and mostly built with V-6, V-8, and V-10 configurations and cylinder bore of 12-1/2 inches. These engines ranged up to 1,100 horsepower, and all ran at a constant speed of 450 rpm.

Many of the Page diesel-powered draglines built in the 1940s and 1950s were equipped with two diesel engines and an electric swing system. Models 621 and 625, in the 6- to 9-cubic-yard class, had a five-cylinder engine on the main deck driving the hoist and drag drums through clutches and brakes, while a three-cylinder engine mounted on an upper deck drove a generator for the electric swing motors.

The diesel-powered Bucyrus-Monighan draglines were usually specified with either Cooper-Bessemer or Fairbanks-Morse slow-revving engines. These ran at about 450 rpm and produced up to 1,000 horsepower in the case of the Model 480-W.

Power to electric-operated draglines is led into the machine via a heavy-duty trailing cable. The incoming AC power feeds constantly running synchronous motors that drive DC generators for each motion. The generators provide power to their respective DC motors through Ward Leonard control. As mentioned in the stripping shovel section, the reliable Ward Leonard control maximizes pull at stall speed, and precisely controls speed and direction of each motion at the operator's command.

WALKING DRAGLINE OPERATION

Although not originally designed for mining, draglines were soon found to be ideal machines to strip overburden in surface mines. If well organized, a dragline operation is the most efficient mining method for nearly all geological conditions encountered in strip mines. Like the stripping shovels described in Part I, a dragline digs in long strips or cuts, and casts the overburden into the neighboring cut from which the mineral has already been removed. But the dragline is much more flexible in its work capability. With a dragline at his disposal, the mining engineer has many options from which to select the best mining plan. Unlike the stripping shovel, which is positioned in the bottom of the cut and thus limited by the amount of material it can dispose of, the walking dragline's low ground pressure allows it to work on top of the face to be excavated. Here its greater range can be used to cast the material further afield, allowing mine operators to make a deeper or wider cut. A dragline can also rehandle material to form a "bridge" of material across the cut. Positioned on this bridge, the dragline can then dump even further out onto the spoil pile.

A dragline can also work on a bench below ground level. The bench is formed by the dragline itself by "chopping" the overburden above the bench level. The material below the dragline elevation is then dug and cast in the usual way. When a dragline is chopping, the bucket is filled as it is pulled down the face toward the dragline. Production in this working mode is acceptable, but not as high as when the bucket is pulled uphill.

The walking dragline operating crew usually consists of three people: the operator, the oiler, and a ground man. In addition to handling machine lubrication, the oiler must keep the machine clean. He also functions as a relief operator, and spends time in the operator's seat so he can become familiar with all the

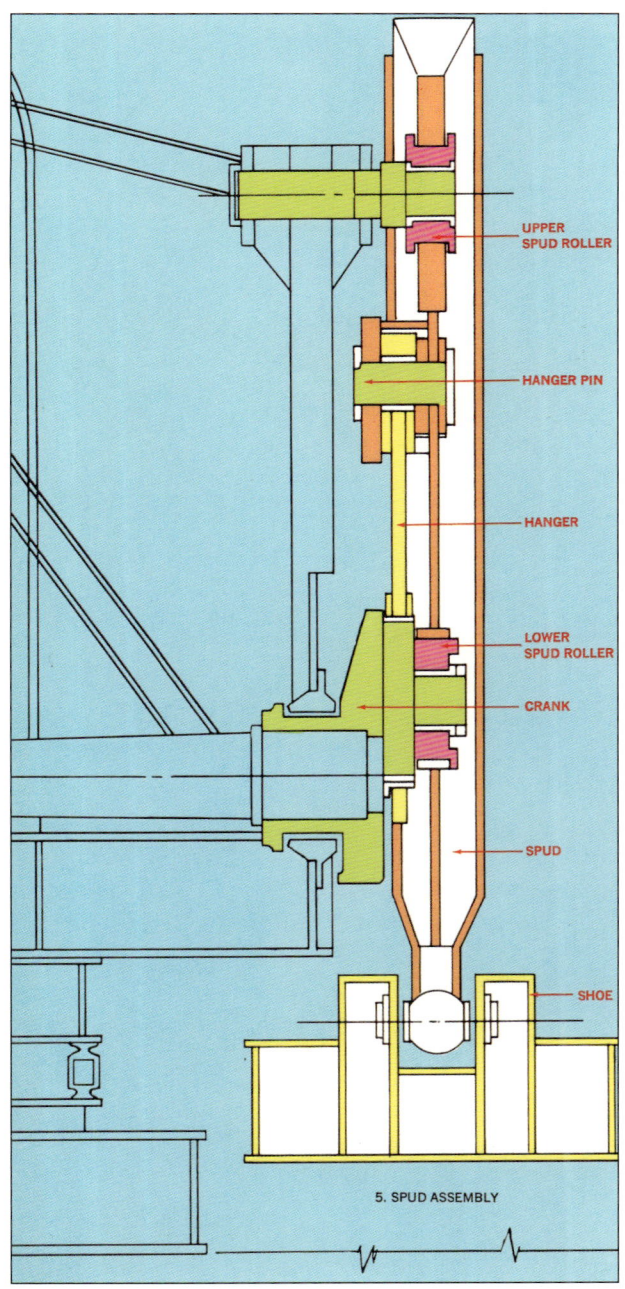

Figure 4.7

Appearing large for a bucket of only 11 cubic yards, this Rapier W600 long-range dragline is uncovering ironstone near Scunthorpe, England, for the British Steel Corporation. It is positioned on the mineral, and chops the overburden from the face at right. The ironstone is being loaded by the electric shovel in the middle distance.

—David R. Wootton

This Rapier W1350 was shipped from the Ipswich Works of Ransomes & Rapier, Ltd., to the Wabamun strip mine west of Edmonton, Alberta, in 1962. The 33-yard machine had a productive life of almost 30 years before being idled, and finally cut up for scrap in 2000.
—Keith Haddock collection

One blast, and the dragline boom comes crashing to the ground. Owned by Miller Mining, the Rapier W600 has finished its working life of over 30 years, stripping coal in the County of Yorkshire, England.
—Peter Grimshaw collection

machine's operations. The oiler is usually the second in command, and next in line as an operator when he becomes fully trained. He follows a maintenance checklist, which includes walking up the long catwalk to the top of the boom at least every other day. If he sees cracks in the boom structure, he reports these to the maintenance crew for repair while they are still minor.

The ground man keeps the machine's power cable out of the way of other vehicles, and clear of the shoes when the dragline is walking. You don't want a 4,000-ton machine stepping on its own power cable! The dragline is fitted with a two-way radio system with stations throughout the machine so that the ground man and oiler can communicate with the operator. The ground man, however, spends most of his time operating a bulldozer assigned to the dragline.

He uses this to keep the dragline working bench absolutely level. Loose rocks or uneven ground, especially in frozen conditions, could cause severe damage to the underside of the tub. The ground man also uses the bulldozer to push back the "roll" caused by the dragline from time to time. The "roll" is the mound of material pulled up by the bucket, and which accumulates near the fairlead—too close to be picked up by the dragline.

The operator controls the machine's digging motions through two hand levers and two foot pedals. The right lever controls bucket hoisting and lowering, and the left lever controls the drag motion. The two foot pedals control the swinging of the machine, right or left—one pedal for each direction. The levers and pedals are arranged so that power is off in their central positions. As the operator moves a lever or pedal away from its central position, it applies power through the DC motors to the motion selected—hoist, drag, etc. The further the operator moves the lever or pedal, the more power is applied. Moving the lever or pedal in the opposite direction

Large numbers of walking draglines have been built in the former Soviet Union (now the Commonwealth of Independent States). This 20-yard ESH 15-90 built by UZTM is the most popular, with 141 recorded as built since 1959. The upper part of the boom is a single tube supported by cables tensioned in all directions.
—Keith Haddock collection

reverses the motion. Some recent draglines have been equipped with "joystick" controls. Since joystick levers move in all directions, one or two functions can be performed with each lever at the same time, and foot pedals are eliminated. Operators favor the joystick controls once they get used to the change, and several machines have been retrofitted with this system. Younger operators readily take to the joystick controls because of their similarity to computer game joysticks.

Since no brakes are used during the digging cycle, the operator needs a lot of skill to synchronize the hoist, drag, and swing motions all at the same time. The masses in motion are so huge that some lag is inevitable between the operator's lever movements and the action indicated. Therefore the operator must anticipate the bucket's movements well in advance.

In addition to the four main operating controls mentioned above, there are many other switches in the operator's cab. There are "parking" brakes for the hoist and drag and swing machinery. These are not used during the digging cycle, but only for holding the drives stationary when power is cut off. Other switches are for the many lights, air conditioners, motor excitation, and emergency stop.

MOVING LARGE WALKING DRAGLINES

Walking draglines, like stripping shovels, have to be moved from one location to another in many small components of roadable size, as described in chapter 1. Thousands of parts are shipped by rail or road from the factory when the machine is new. This may take upward of 250 railroad cars. Erection on site may take a year or more, under the supervision of a manufacturer's representative.

When a dragline is moved from one location to another, it must be disassembled and moved in small

Popular dragline in the former Soviet Union is this Esh 20.90. Like other UZTM machines, it features a hydraulic walking system consisting of two hydraulic cylinders connected to each shoe. The predetermined sequence of oil pumped in and out of the hydraulic cylinders imparts a circular motion to the shoes, similar to the motion produced by the more usual mechanical systems.

—Keith Haddock collection

components to the new site. The long duration of most mining operations makes the high cost of dismantling and erecting these giant machines insignificant when averaged over their working life, which can extend to 30 years or more. Some machines have been moved five or more times during their working life. More often, though, a machine will spend its entire life at one location, where it is scrapped upon job completion.

Because of the high cost of dismantling these complex machines, operators look for alternative ways of moving the machine when the distance is not too great. In many cases, mine operators have walked the draglines great distances—20 miles or more—under their own power. A "walk" in this fashion may take many months of planning. Railroads, highways, rivers, power lines, underground cables, and pipelines all pose challenges. Miners must make arrangements with local county or highway departments to temporarily close roads for about a day, to allow the machine to cross. The roads themselves must be protected by a layer of earth, at least 6 feet thick, to minimize the pressure exerted as the tremendous weight of the machine passes over the road surface. Power lines must be isolated, then cut and laid on the ground. This is done by the power company, which must make advance arrangements with customers who may be affected. River crossings require culverts and lots of competent fill material to ensure the machine doesn't become stuck on its approach to, or crossing, the river.

Another method of moving large walking draglines has recently proved successful. Lampson International Ltd. of Kennewick, Washington, has developed some two-crawler transporters capable of carrying up to 4,000 tons each. Four of these are able to carry giant walking draglines in one piece! Recently, draglines have

Not all walking draglines come in big packages. This prototype Terasmies 3 walking dragline, built by Locomo of Finland, has a capacity of 1/2 cubic yard.

—Keith Haddock

successfully been moved by this method in the United States and Australia.

CURRENT WALKING DRAGLINE SITUATION

In the 1980s the dragline market became somewhat saturated, and new machine orders dropped off dramatically. The inherent long life of walking draglines largely contributed to this sales decrease. Those in coal mines were of sufficient capacity to meet long-term contracts, and were designed to meet a certain constant demand by power generating stations. A dragline is designed to last from 20 to 30 years at least, but with major rebuilds, some 50-year-old machines have continued to operate profitably. The investment in these machines is so high—$15 million to $40 million—that owners rebuild and modify components rather than buy a new machine. In other words, if something breaks, fix it. If a gear, motor, or bearing wears out, replace it. Excessive wear or malfunction rarely requires scrapping an entire machine. Most machines are scrapped because of lost contracts or lack of further work.

Finally, there has been a drastic shift from coal production in the Midwest, where most of the large stripping machines operated. Although coal production is increasing on a yearly basis in the United States, and has done so throughout the 1990s, that increase is coming from the western states, particularly Wyoming. Here, rich coal seams lie under relatively shallow overburden, which can be moved efficiently by trucks and shovels. Western operators have installed a few big draglines in recent years as, even here, overburden depth is increasing.

In order to comply with the U.S. Clean Air Act, several midwestern power generation companies have found it cheaper to ship low-sulfur coal from the west than to convert their power stations to burn the local, high-sulfur coal.

As the mines in the west experience higher mining ratios (overburden moved per ton of coal mined), and some of the older mines are worked out, manufacturers expect mine operators to order new draglines.

CHAPTER FIVE

Walking Dragline Chronology

The only one of its kind, the 8-yard Page 631 dragline is shown near Hazleton, Pennsylvania, after being idle for several years. One of the company's 600 series machines launched in 1935, it features Page's improved walking system.
—Keith Haddock

John W. Page is acknowledged as the builder of the world's first dragline machine, in 1904. He was a partner in the contracting firm Page & Schnable, which needed a machine to dig below grade level. He devised a crude wooden machine with a swinging boom controlled by cables, carrying a bucket to hold about 1 cubic yard of material. The steam-powered machine worked well, so Page built more machines for his own use.

Oscar Martinson, chief engineer of the Monighan Machine Company, invented the first walking device for a dragline in 1913. Known as the "Martinson Tractor," it was first fitted to a Model 1-T.
—Keith Haddock collection

It wasn't long before other contractors asked Page to build similar machines for them, and he was keen to oblige. Eventually, building draglines proved more profitable than contracting, and Page Engineering Company was incorporated in 1912 in Chicago to build draglines and dragline buckets.

A number of other manufacturers entered the dragline business between 1908 and 1912, including Monighan and Heyworth-Newman in the Chicago area. Monighan and Page built some machines in partnership prior to going their separate ways (see Appendix 1). Heyworth-Newman was purchased by the Bucyrus company in 1910, and the following year Bucyrus offered a new line of draglines. The first of these was the Bucyrus Class 24, a steam monster carrying a 3-1/2-cubic-yard bucket on a 100-foot boom. It was the world's largest dragline up to that time, and many were sold from 1911 to 1930. The last Class 24, and probably the oldest known dragline still existing, is preserved as a nonworking exhibit at the Reynolds Alberta Museum, Wetaskiwin, Alberta, Canada. This steam-powered example, built in 1917, is mounted on skids and rollers.

In the early years of dragline bucket development, a number of manufacturers patented different designs. The Browning Scraper Bucket required two hoist ropes as well as the drag rope. The Heyworth-Newman design employed a three-part hoist line with a sheave on the bucket arch. The Monighan, Austin, and Iverson types featured a latch lever mechanism that held the bucket horizontally until a sharp pull on the drag rope released the latch, and allowed the bucket to dump vertically. An extra advantage of this latter bucket type was that it could be hoisted at any point under the boom without dumping its contents.

But the original, the bucket that Page patented in 1904, was the one that eventually overruled all others in the field. The Page design requires one hoist rope, one drag rope, and no latch mechanism. A short dump rope running from the bucket arch, over a sheave at the hoist rope anchor, and then down to the drag rope anchor, keeps the bucket horizontal as long as there is tension between the hoist and drag ropes. As the drag rope is paid out, drag tension is decreased until the bucket reaches a position approximately under the boom point when it dumps automatically. This simple, but effective bucket system proved superior to all other designs, and since the early 1920s has been adopted by all dragline bucket manufacturers.

None of the draglines up to 1912 were self-propelling, so they were cumbersome to move. They were mounted on rails, or skids and rollers. To move, the rail wheels or rollers were released, the bucket anchored in the ground ahead, and the dragline rolled itself toward the bucket using its drag rope. Laying rails or skids and rollers made this system complicated and slow. Operators wanted some means of self-propulsion for the machines.

Page Engineering Company patented this complicated four-legged walking device for its draglines about 1925. Shown on an 8-yard Model 430, the system was operated by chains, racks, and pinions.
—Keith Haddock collection

In 1912, Bucyrus fitted a Class 14 dragline with self-propelling crawler tracks. It was the first excavator of any type to be so mounted. Although the crawlers worked well on this 2-yard machine, and were ideal for smaller revolving shovels, crawler tracks resulted in ground pressures too high to support the giant draglines on the soft ground where most of them worked. These machines needed a different way to move.

THE FIRST WALKING DRAGLINES

In 1913, Oscar Martinson, an engineer with the Monighan Machine Company, invented the radical idea of attaching two movable shoes, one on each side of the dragline's revolving frame. Martinson's innovation changed dragline mobility forever. The first "walking" device, known as the Martinson Tractor, was fitted to Monighan 1-T and 3-T (1-yard and 3-yard) draglines in 1913.

The simplicity of Monighan's walking system made it so successful. Each shoe is suspended by chains from a beam, which in turn is hung from an eccentric trunnion. One trunnion is fixed to each end of a long horizontal walk shaft, which runs across the machine. To take a step, the walk shaft rotates the trunnions in a circular motion so that both shoes touch the ground at the same time. Further rotation lifts the leading edge of the dragline's circular tub off the ground, pulls it ahead the distance of one step,

and then lowers the machine gently back to the ground. The shoes continue to rotate and the process is repeated step by step. Changing direction (steering) is just a matter of pointing the rear of the machine in the desired direction when the shoes are off the ground. When digging, the dragline sits on its circular base or tub and the shoes hang from the revolving

continued on page 84

A close-up view of Page's first walking system. Attached to an independent frame, the vertical legs were powered by rack and pinion. During walking, the entire base was lifted off the ground as shown, and then moved backward by a chain and gear system supported by the frame.
—Keith Haddock collection

Above: Martinson improved his waking device in 1925 and launched the Monighan 3-W, with a capacity of 3 cubic yards. The solid oval path for the eccentric wheel resulted in a positive reliable motion. This is a late Model 3-W salvaging river gravel in a Basalt Rock gravel operation.

—Bucyrus International, Inc.

Below: A typical Bucyrus-Monighan walker of the 1930s, this diesel-powered 6150 is working near Larksville, Pennsylvania. The bucket holds 6 cubic yards swinging from a 150-foot boom. The peaked roof of the house distinguishes Monighan draglines of this era.

—Walter Bennett collection

Above: The world's largest dragline in 1932 was this Bucyrus-Erie 950-B. Swinging a 12-yard bucket, the electric machine also boasted the world's longest boom, at 250 feet. Shipped to a limestone quarry in Brazil, it was the only machine of this type.

—Bucyrus International Inc.

Below: One of the most popular walking draglines was the Bucyrus-Monighan 5-yard 5-W, launched in 1935. A total of 141 5-Ws were built up to 1971, in either Chicago, Illinois, or Lincoln, England. The one in the picture is a Ruston-Bucyrus.

—Keith Haddock collection

The Bucyrus-Monighan 15-W was one of the largest walking dragline to feature winding drums operated by clutches and brakes. Only three 15-Ws were built, all electrically powered. The machine in the picture worked at Manalta Coal Ltd.'s Vesta Mine in Alberta, with a 12-yard bucket on a 215-foot boom.
—Keith Haddock

The British firm of Ransomes & Rapier Ltd., first entered the walking dragline business with this W170 in 1938. It carried a 4-yard bucket on a 135-foot boom. The W170 shown here is uncovering ironstone in central England, where it worked until it was scrapped in 1971.
—Keith Haddock

Continued from page 81

frame at the side of the machine. The tub is made of such large diameter that ground pressure is reduced to a minimum.

The walking device was so successful that Monighan developed a line of seven machines in 1/2-yard increments from the 1-yard Model 1-T, up to the 4-yard Model 4-T. The 1-T became the second-most popular walking dragline of all time in terms of numbers sold. The company shipped a total of 117 from the Chicago factory up to 1926.

With the invention of Martinson's walking device in 1913, Monighan's draglines gained a significant advantage over the Page machines, and several years would pass before Page perfected a walking device. The company was not as aggressive as Monighan, and found an adequate and satisfactory business in supplying basic rail-mounted draglines to well-established customers. Page obtained many patents for his designs, including a cableway excavating machine in 1916, and several dragline bucket improvements.

Page draglines began to appear with a crude walking system about 1923. In contrast to the simplicity of the Monighan design, Page came out with a very complicated system consisting of three vertical legs, two in front and one at the rear. The legs were attached to an independent heavy frame structure enclosing the dragline's tub and machinery house, and were operated by rack and pinion in a complicated mechanical system of spur gears, shafts, and chains. To walk, the dragline's three legs were extended downward to lift the entire machine clear of the ground. Once raised, the machine was pulled by a chain and sprocket system along a horizontal roller path inside the heavy frame. To complete one step, the legs lowered the machine back to the ground, and the chain pulled the frame back along the roller path to the starting position.

Page did not produce a large number of draglines with this walking system. The most popular of this design was the 411W of 2-1/2 yards capacity. The first digit of the Page model numbers denotes the design series, while the last two numbers denote the diameter of the swing circle in feet. About 15 of the Model 411W were built between 1923 and 1930. And a single 8-yard monster, the Model 430 with four walking legs, was built in 1930. In addition to these walking models, Page continued to market and sell nonwalking machines well after this period. There was very little activity from the company during the Great Depression. Then in 1935, Page unveiled the first of its 600 series walkers, the Model 620. This featured a vastly improved walking system that was utilized on Page draglines until the demise

The 2-1/2-yard Rapier W90 appeared in 1943, offered with either diesel or electric power. This W90, however, swings a 1-1/2-yard bucket on a special boom 130-feet long. It is working in 1985 for the London Brick Company near Peterborough, England.
—Keith Haddock

A fleet of eight Rapier W150 walking draglines was operated by the London Brick Company at its various clay pits in eastern England. The 6-yard W150 was in production from 1944 until 1963.
—David R. Wootton

This Model 7200 is the first walking dragline built by Marion. Shipped in 1939, it was still working some 55 years later. The picture was taken at the American Aggregates' gravel pits near Indianapolis, Indiana. The 7200 normally carries a 7-yard bucket.
—Keith Haddock

of the company in 1988. A little more complicated than the Monighan system, it used eccentric drive to the shoes, which were attached to the machine by walking spuds or legs. This walking system is described in the previous chapter.

In 1925, Martinson hit again with an improved walking system that eliminated the suspension chains by substituting a cam wheel running in an oval track in a frame pivoted to the shoes. The new 'W' series machines went into production starting with the first 3-W (with a 3-cubic-yard bucket) in 1925, and immediately made the existing 'T' series machines obsolete. To Martinson's credit, Bucyrus-Erie retained his walking system when it took over the Monighan company in the early 1930s, and continued to use it right up to the present day. The number preceding the 'T' or 'W' on the Monighan machines denotes the standard bucket size of the model in cubic yards.

The most popular Marion walking dragline was the 7400. Over 90 were sold during a production run lasting from 1940 to 1974. In the 11-yard class, the 7400 was available with diesel or electric power. The machine shown is stripping coal at Forestburg, Alberta, Canada. Well traveled, this machine was originally shipped new to Australia in 1948 for a short-term job. Then it worked in South Wales in Great Britain, for about 20 years until going to Canada in 1974.
—Keith Haddock

The largest of Marion's first three draglines was the 7800, launched in 1942. With a 30-cubic-yard bucket on a 185-foot boom, it took the title of the world's largest dragline. This one is at work at Peabody Coal Company's River King Mine, Illinois.

—Keith Haddock

With the new walking system successfully established, Monighan went on to build some record-beating draglines. The 6-yard 6-W appeared in 1926 with a standard 100-foot boom, followed by the 6150 in 1929 with a 150-foot boom. The company crossed the 10-cubic-yard threshold in 1934 with the launch of the 10-W. Most of these Monighans were sold as diesel machines with the drums operated through clutches and brakes. Nearly all were equipped with Fairbanks-Morse or Cooper-Bessemer slow-revving marine-type engines (350 to 450 rpm). The four-digit model series (6150, 6160, and 8160) denoted the standard bucket size and boom length, i.e., the 8160 swung an 8-yard bucket on a 160-foot boom. It is not clear why these three models were so designated because they were marketed simultaneously with the similarly designed 'W' series draglines in the same capacity range.

Bucyrus-Erie Company took a controlling interest in the Monighan company in 1932 and changed its name to the Bucyrus-Monighan Company. This gave Bucyrus-Erie access to all the walking dragline designs. Bucyrus-Erie promptly launched the 950-B, the largest dragline built up to that time. While the Bucyrus-Erie 950-B established new standards and lasting fame as a stripping shovel (see chapter 2), it was introduced as a dragline that swung a 12-yard bucket on a 250-foot boom, the longest in the world. Its operating weight was just over 1,000 tons. Shipped to Brazil in 1935, the dragline was used in a large cement quarrying operation. The 950-B was regarded as an engineering masterpiece, and it laid the foundation for much larger Bucyrus-Erie machines that appeared over the next three decades. Although the manufacturer's literature stated the 950-B could easily be converted from dragline to shovel, no such conversion ever took

Above: When Bucyrus-Erie launched its 1150-B in 1942, it was the company's largest dragline at that time. Carrying a 20-yard bucket on a 200-foot boom, it had an operating weight of 1,265 tons. The 1150-B shown is loading material into a portable crusher, which is discharging onto a conveyor. The location is the South Agnew Mine in Minnesota, and the date is 1949.

—Bucyrus International

Below: Distinguished by its unique two-part boom, the Bucyrus-Erie 1250-W swung the world's largest dragline bucket in 1960. The machine shown is the Anthracite King, uncovering anthracite near Hazleton, Pennsylvania, with its 35-yard bucket and 225-foot boom.

—Bucyrus International Inc.

The 1260-W turned out to be one of the Bucyrus-Erie's most popular large draglines. Thirty-three were shipped between 1965 and 1992. This one, belonging to K & J Coal Company, is working near Westover, Pennsylvania.

—Bucyrus International Inc.

place. A total of 11 950-Bs were built, but all except the first unit previously mentioned were sold as stripping shovels.

Bucyrus-Erie left the former Monighan plant in Chicago intact after buying the company, and also gave it independence from Bucyrus-Erie's South Milwaukee management in initiating new designs. A prime example was the popular Bucyrus-Monighan 5-W, unveiled in 1935 as a 5-yard dragline carrying a 120-foot boom. Available with either diesel or electric power, the 5-W was popular in surface mines, gravel pits, irrigation projects, and all types of construction work. It still holds the all-time best-selling record for a walking dragline. The Monighan plant built a total of 79 of these popular draglines, and a further 62 were built in the Ruston-Bucyrus plant at Lincoln, England, up to 1971.

Just prior to World War II, the Monighan plant added the very popular 7-W and 9-W machines to the range. After the war, further new machines were introduced, including the first 200-W (6 yards) in 1945, and the 500-W (12 yards) in 1946. These machines, and the big diesel-powered 450-W and the 480-W (14 yards), kept the Chicago plant busy until its closure in 1958. The last machine built in the Chicago plant was a 7-W shipped early that year. Walking dragline production was then transferred to South Milwaukee.

ENTER RAPIER AND MARION

In the late 1930s, two other walking dragline names emerged—Rapier and Marion. The British firm of Ransomes & Rapier Ltd. designed its first dragline in 1938, the Rapier W170, and put it to work the following year. It carried a 4-yard bucket on a 135-foot boom, and featured the patented Cameron & Heath walking system, in which the shoes were attached to an eccentric cam running in a roller bearing. The 2-yard W-80 followed in 1940, then the 2-1/2-yard W-90 in 1943. Some of these machines, with long booms (130 feet) carrying 1-1/2-yard buckets, were used in brickworks

The world's largest walking dragline in 1961 was the 40-yard Rapier W1800. The one pictured is working at the Boundary Dam Mine of Luscar Ltd., near Estevan, Saskatchewan, Canada. At the time of writing this machine was still operational.
—Luscar Ltd.

clay pits. The following year, Rapier introduced its popular 6-yard W150, which remained in production until 1963. The biggest customer for this model was the London Brick Company, which eventually operated eight at its clay pits in eastern England. At the time of writing, at least one is still operating.

The Marion Steam Shovel Company entered the walking dragline market in 1939 with the diesel- or electric-powered Model 7200, swinging a 5-yard bucket on a 120-foot boom. Later, 7200s carried 7-yard buckets. The following year, Marion came out with the 10- to 11-yard 7400. And then came the big one! In 1942, Marion introduced the world's largest dragline, capable of swinging a 30-yard bucket on a 185-foot boom, and weighing a massive 1,250 tons. Launching the 7800 was an impressive achievement for Marion, which had only been in the dragline business for three years. The 7200, 7400, and 7800 were three extremely successful draglines that Marion retained in its product line for over 20 years. Here was a classic case of getting it right the first time. The company sold more than 90 Model 7400s from 1940 to 1974—an incredible manufacturing life of some 34 years! The first Marion 7200 was still working some 55 years after it was built, and several Marion 7800s are still in service at the time of writing.

Launched in 1944, the Bucyrus-Erie 1150-B competed directly against the Marion's 7800. A dragline of similar weight and capacity to the 7800, the 1150-B was developed from the 950-B, which had appeared almost a decade earlier. Its circular tub measured 44 feet in diameter, and the machine was offered with booms from 180 to 215 feet long. Electrical equipment included two hoist motors and two drag motors, all of 425 horsepower, plus three 125-horsepower swing motors and two 100-horsepower propel motors.

Sixteen 1150-Bs were delivered in the United States prior to 1950. After only a few years' work,

four of these were dismantled and shipped to the United Kingdom following World War II as part of the postwar recovery program. There, the machines assisted with the rapid recovery of much-needed coal. All four machines had a long life in their new home, working well into the 1980s. One has been preserved by a private preservation society, the St. Aidan's Trust, at a mining interpretive site near Leeds, Yorkshire. At the time of writing, this 1150-B is the world's only preserved walking dragline. In 1950 the 17th and last new 1150-B also went to the United Kingdom to strip overburden from iron ore deposits at Corby, Northamptonshire, where it worked until 1980.

DRAGLINE SIZE INCREASES

In 1951, Bucyrus-Erie upgraded the 1150-B to the 1250-B, a dragline with more power and capacity than its forerunner. Power was increased to 500 horsepower for its two hoist and two drag motors, and 150 horsepower for its three swing motors. Booms up to 235 feet, and buckets up to 33 cubic yards were offered. After shipping eight 1250-Bs, Bucyrus-Erie upgraded its largest dragline again in 1959 and designated it the 1250-W. Although gaining a modest increase in capacity to 35 yards, and with booms up to 245 feet the 1250-W's most distinguishing feature was its boom design. It had a unique two-part boom suspended by ropes from a mast pivoted on the lower boom section. The idea was to facilitate changing boom lengths by adding different-length top sections. But to the author's knowledge, no such boom changes ever took place on the six 1250-Ws in the field. Four of these worked in the phosphate fields of Florida, while the other two stripped coal in Pennsylvania. Rounding out the 1200 series draglines was the modern-looking 1260-W, which first appeared in 1965. With buckets ranging from 30 to 42 yards, this dragline was fitted with a computer-designed triangular boom. It had only three main chord members instead of the usual four. The 1260-W turned out to be one of Bucyrus-Erie's

One of only two Bucyrus-Erie 1550-W draglines built is operating at Consol's Burning Star No .4 Mine near Pinckneyville, Illinois. The 65-yard machine is joined in the same pit by a Marion 5761 stripping shovel.
—Keith Haddock collection

most popular draglines, with 33 being sold up to 1990. The last one went to work in 1992 for NB Coal Ltd. in New Brunswick, Canada.

Ransomes & Rapier Ltd. claimed title to the world's largest dragline when its W1400 went to work in 1951. Although its 20-yard bucket had been exceeded earlier, the Rapier W1400's 282-foot-long boom, and its operating weight of 1,880 tons, were much greater than any dragline built up to that time. The W1400 was the first dragline to be fitted with a boom made of tubular members. These were filled with compressed gas and the operator monitored the

Left: The 1570-W first appeared in 1973 as a 70-yard class machine. Still current, the model has received many upgrades, boosting its capacity to 80 cubic yards. The machine in the picture has an extra long boom measuring 345 feet, with a corresponding bucket reduction to 58 cubic yards. It is working at Luscar's Paintearth Mine in Alberta, Canada.

—Keith Haddock

Peabody Coal Company purchased the only two Marion 8900s in 1967. The one pictured is swinging a 155-yard bucket at the Hawthorne Mine, Indiana. It initially worked at the Dugger Mine with a 145-yard bucket. The 8900 was the second largest dragline built by Marion.

—Eric C. Orlemann

pressure level from the cab. If a crack occurred, the operator would recognize the drop in pressure and the machine would receive required service. The cantilever-type triangular-shaped boom is instantly recognized as a unique Rapier feature. Three of the W1400 machines worked in the ironstone fields of central England.

In 1961, Rapier returned with an even larger machine, the W1800, reestablishing the company as the builder of the world's largest dragline. Weighing over 2,000 tons and carrying a bucket of 40 cubic yards, the first W1800 went to a large open pit coal mine in South Wales operated by contractors George Wimpey & Company. Other W1800s worked on British ironstone stripping. One is still operating at Luscar Ltd.'s Boundary Dam Mine in Saskatchewan, Canada.

At its factory in Ipswich, England, Ransomes & Rapier boasted a large shop designed specially to assemble walking draglines. The concrete floor was level to within 5/32 of 1 inch, so that components could be laid out on the floor and welded together without additional leveling or shims. The company fully erected some of its draglines for export in this shop, including machinery house panels, before dismantling, shipping, and reerecting the machines at the work site. It seems that Rapier wanted to be absolutely sure that everything would fit at the erection site!

Two years after the W1800 went to work, Bucyrus-Erie delivered its first 1450-W. With an operating weight of 2,916 tons, it broke the world size record again. A total of six 1450-Ws were sent to the Midwest for coal stripping, all equipped with 250-foot booms. Based on the 1450-W design, two 1550-Ws were sold in 1968. One went to Consolidation Coal's Burning Star Number 4 Mine in Illinois, and the other was shipped to England, where it became the famous *Big Geordie*, operating at Derek Crouch's Radar North opencast coal site. Both 1550-Ws swung 65-yard buckets. The 1550-W was upgraded to the 1570-W, a still-current model, which has received many upgrades since the first was shipped in 1973. At the time of writing, 46 1570-W draglines have been shipped, with buckets ranging from 58 to 80 cubic yards depending on boom length.

In 1993, Marion shipped two Model 8750s with 420-foot-long booms, a length never exceeded. One went to Callide Coalfields in Australia, the other to Fording Coal Ltd. (pictured) for its Genesee Mine, near Edmonton, Alberta. This one carries a 106-yard bucket.
—Fording Coal Ltd.

Marion shipped its largest-ever dragline to Amax Coal Company in 1973. The 150-yard 8950 weighed 7,300 tons, and spent its working life at the Ayrshire Mine in Indiana.
—Eric C. Orlemann collection

In 1963, Marion startled the dragline world when it introduced its Model 8800. This world record beater represented a massive jump in size, with an 85-yard bucket on a 275-foot boom. For the 8800, Marion had to redesign its traditional single-crank walking system to a two-crank system to support the model's 6,000-ton weight. The 8800 is described further in chapter 6, and its walking system is described in chapter 4. Peabody Coal Company purchased the one and only 8800, which was later upgraded to a 100-yard machine.

Bigger draglines followed in the 1960s, the decade of the big strippers. Marion shipped two Model 8700s, in 1963 and 1965. The first swung a 70-yard bucket on a 225-foot boom, and went to work for Peabody Coal in Ohio. The second went to Pagnotti Enterprises, who still use it to dig anthracite from a deep open pit in eastern Pennsylvania. It carried an 85-yard bucket on a 300-foot boom, the longest in the world at that time.

In 1964, Bucyrus-Erie introduced the first of four 2550-Ws, two of which were purchased by Amax Coal Company for stripping coal in Indiana. The 75-yard 2550-Ws were equipped with booms of 275 or 300 feet, and featured a new walking system known as the "cam and slide" system. It was designed by Bucyrus-Erie for machines greater than the 70-yard class, and is described in chapter 4. The 2550-W grew into the

This Bucyrus-Erie 2570-W is uncovering coal at Amax Coal's Chinook Mine, Indiana. Introduced in 1973 as an upgrade to the earlier 2560-W Model, the 100-yard class 2570-W was one of B-E's most successful draglines, with 27 sold worldwide.
—Keith Haddock collection

2560-W, two of which were sold to Peabody Coal company in 1969. The first of these swung an 85-yard bucket on a 295-foot boom. It worked at the Elm Mine, Illinois, which was eventually owned by Midland Coal Company.

Peabody Coal purchased the only two Marion 8900s in 1967, one for work in Australia, the other for a coal mine at Dugger, Indiana. These carried 130- and 145-yard buckets, although the Indiana machine was upgraded to a 155-yard bucket in 1993. They were designed on similar lines to the earlier 8800, and utilized the double-cam walking system.

Then came Bucyrus-Erie's "*Big Muskie*," the most famous of all draglines. Built late in the decade, this machine was able to capture and keep for all time the title of the world's largest-ever dragline. The enormous 4250-W, its technical name, also used the largest

Amax Coal Company purchased the only two Bucyrus-Erie 3270-W draglines in 1977. This one makes overburden work easy at the Delta Mine in Illinois. Both machines were fitted with 176-yard buckets and 330-foot booms. They are still the two second-largest draglines ever built.

—Bucyrus International Inc.

bucket ever to swing from an excavator, 220 cubic yards. *Big Muskie*'s story is told in chapter 6.

When *Big Muskie* started work in 1969 at the Muskingum Mine of the Central Ohio Coal Company, her vast size was just the latest on a rapidly ascending curve. Ten years earlier, no one had contemplated a machine of such gigantic proportions. Manufacturers predicted the upward curve would continue, and went on designing even larger machines. Marion designed its 9600

Erection of the 3270-W dragline at the Delta Mine in 1977. A bevy of heavy-duty cranes, including a 300-ton Lima 7707, right, were needed to erect the world's second-largest dragline.
—Keith Haddock collection

The operator of the 3270-W can be seen in one of the two cabs on this giant machine. He is putting the machine through its paces, shortly after erection was completed in 1979 at Amax Coal's Delta Mine in Illinois.
—Bucyrus International Inc.

in a bid to compete with the 4250-W, and designers talked of 1,000-yard machines, but nothing larger than *Big Muskie* was ever constructed. As explained in chapter 4, a new generation of more-reliable, technically improved draglines appeared in the 1970s which, although smaller, could move overburden at lower unit cost.

MODERN DRAGLINES FOR THE COAL BOOM

In 1971, Marion brought out its 8750. This was the first of a redesigned line of highly efficient 8000 series draglines, initially for buckets from 80 to 115 cubic yards, and operating weights up to 7,100 tons, depending on the boom and bucket configuration. The walking device on the new 8000 series draglines featured a single-cam system incorporating

Ransomes & Rapier launched the W2000 dragline in 1977. The 45-yard unit featured a conventional boom, departing from the cantilever-type boom design found on Rapier's earlier models. Today, these machines are sold by Bucyrus Europe Ltd. The machine pictured is removing overburden at RJB Mining's St. Aidan's site near Leeds, England.

—David R. Wootton

an eccentric cam with an outboard bearing support. This meant that both ends of the walk shaft on each side of the machine were supported by the dragline's frame. The 8750 turned out to be one of Marion's most successful large draglines, with 24 sold up to 1993.

A new idea in walking dragline design—the modular concept—was introduced in the 1970s. Bucyrus-Erie was first with this 11-yard 380-W in 1979. Modules were fabricated in the plant, and then bolted together at the work site. Erection and relocation times were reduced from many months to just a few weeks.
—Keith Haddock

Although identified with the same 8750 model designation, the later machines bore almost no resemblance to the early 8750s. As each new machine was commissioned, Marion incorporated design changes and improvements. The last 8750, shipped to Fording Coal's Genesee Mine, Alberta, Canada, in 1993 has no less than 16 swing motors mounted in the revolving frame below deck level. It swings a 106-yard bucket on a boom 420 feet long, the longest ever installed on a dragline. The 8750 shipped in 1991 to Curragh Queensland Mining in Australia carries a 135-yard bucket on a 360-foot boom.

Designed on similar lines to the 8750, the 60-yard Model 8050 was launched by Marion in 1972. Over the next 14 years, 35 units of this very successful machine were sold, including 20 shipped to Australia. Then Marion introduced two further significant machines in 1973, the 8200 and 8950. The first 8200 was sold to Pittsburg & Midway Coal Mining Company for work in Kansas. Then 33 more were sold, until the last in 1996. Like the long-running 8750, the 8200's design was upgraded many times during its long manufacturing life, so that by the time the last one was shipped in 1996, it was a totally different machine. The 8200 was initially designed as a 65- to 75-yard machine, while the updated version shipped to BHP Coal in 1993 carried a 94-yard bucket.

In 1973, Amax Coal Company purchased Marion's largest-ever dragline, the 150-yard 8950. It was the only one built, and it worked for some 20 years stripping coal at the Ayrshire Mine, Indiana. Its boom was 310 feet long, and the machine's operating weight was 7,300 tons.

Marion's version of the modular dragline was the 7450 in the 11- to 14-yard range. This electric-powered version is excavating anthracite in eastern Pennsylvania for Blaschak Coal Corporation.

—Keith Haddock

To compete with Marion's 8750 in 1971, Bucyrus-Erie lanched its 2570-W, the first of which was purchased by Old Ben Coal Company for use at its mines in southern Indiana. The 2570-W was basically an upgrade from the 2560-W, and employed the same cam-and-slide walking system. To date, 27 of this big dragline model have been sold, with buckets ranging from 80 to 115 cubic yards, and booms up to 400 feet. Like other long-running machines, the 2570-W boasts many product improvements developed over the years.

Another huge dragline was the 3270-W, designed on modern principles by Bucyrus-Erie. Two of these were purchased simultaneously by Amax Coal Company, which put them to work within weeks of each other in 1979. One worked with the Marion 8950 at the Ayrshire Mine, Indiana, and the other at the company's Delta Mine in southern Illinois. They are still the two second-largest draglines in the world with their 176-yard buckets, 330-foot booms, and operating weight of 8,718 tons. The 3270-W employs an eccentric cam and roller bearing walking system, similar to those used on Marion and Rapier machines.

The worldwide rush to coal mining in the 1970s, and a change in management, prompted Ransomes & Rapier Ltd. to announce its reentry in the dragline business in 1975, after a hiatus of some 13 years (See Appendix 1). It came out with a redesigned line of machines, but only the W700 and W2000 series machines were built, the former carrying a standard bucket of 14 yards, and the latter carrying buckets from 35 to 44 yards, depending on the material. The W700 was available with either diesel or electric power. The new line represented a departure for Rapier, in that a cable-supported conventional boom

Another Marion 7450 dragline uncovers coal at one of Willowbrook's mining jobs in 1985. This one carries a 14-yard bucket and is diesel powered.

—Keith Haddock

was utilized, instead of the earlier cantilever type. Several of these new Rapier models went to work in the United Kingdom and the United States, and even greater numbers were put to work by Coal India at various sites in that country.

Meanwhile in the former Soviet Union, walking dragline development has continued apace since 1948, when Novo-Kramatorsky Mashinostroitelny Zavod (NKMZ) built its first machine at the Donetsk plant in the Ukraine. NKMZ has built many walking draglines since then, but none to date have been bigger than 26 cubic yards capacity. On the other hand, the Russian company Uralmash(UTZM) has built some very large machines. Since production started in 1950, 243 walking draglines have left the Uralmash factory in Yekaterinburg. The largest was Model Esh 100.100, built in 1976, which was equipped with a 130-yard bucket. Up to 1976, several of the Uralmash draglines featured a rope-suspended truss boom with a single tubular member for its upper section.

Since then, all Uralmash draglines have been equipped with triangular booms with rigid suspension members. The walking systems on all but the smallest models are hydraulically powered. Two hydraulic cylinders, operating through an electro-hydraulic automatic system, power each walking shoe that moves in an elliptical path. Most Uralmash draglines have operated in the Soviet Union, but a few were exported to Korea, Mongolia, and India. The lastest model is the 32-yard Esh 25.90, which is undergoing tests at the time of this writing.

In 1987, Bucyrus Europe, an offshoot of Bucyrus International, purchased the manufacturing rights to Rapier walking draglines. Bucyrus Europe continued to offer the W2000 Rapier-designed model. At the time of writing, the fleet of W2000s operated by

The W700 was a 14-yard modular dragline introduced by Ransomes & Rapier in 1981. The picture shows one of two electric machines operating at Northern Strip Mining Ltd.'s Godkin Site in central England.
—Keith Haddock

Coal India had grown to 12 units, and in 1999 they ordered three more W2000 draglines for delivery over the following two years.

Not all dragline innovations of the past three decades involved large size. With the rapid increase in surface-mined coal in the 1970s, operators needed a machine that could quickly be dismantled and moved to a new location. Draglines taking many months to move and erect were not suitable for small pockets of coal and short-term contracts. So the modular dragline was born. Bucyrus-Erie was first with its 380-W in 1978, followed by Marion's 7450 in 1980, and Rapier's W700 in 1981. These machines ranged from 10 to 16 yards, and were offered with diesel or electric power. They were built in modular units, designed to be bolted instead of welded together. They could be dismantled and reerected in a matter of a few weeks, and the diesel versions did not have to wait for a power source to be installed. These smaller machines fit mining needs perfectly, and many were delivered up to 1988 when the coal market declined.

Bucyrus-Erie had plans to expand the modular dragline concept into larger machines, and did design work on units up to the 40-yard class. The softened coal market stifled development, however, and only one further modular model was produced. This was the modern, computer-designed 680-W, with buckets from 20 to 25 yards depending on boom length. Five 680-Ws were sold from 1982 to 1988.

The largest modular dragline is Bucyrus-Erie's 680-W, first sold in 1982. The one shown here went to work in 1988 near Cadiz, Ohio. It carries a 25-yard bucket on a 225-foot boom.

—Keith Haddock

Page Engineering Company never produced a supersized record-breaking machine, but its draglines in the small and medium size ranges found a respectable market share. Specializing only in draglines and dragline buckets, Page produced a steady stream of walking draglines in its Chicago plant. The 621 and 625 series diesel-powered machines, featuring Page's own low-revving diesel engines, were very popular in the 1940s. In a typical installation, two diesel engines running at 450 rpm were utilized. One (five-cylinder, 550-horsepower engine) powered the hoist and drag functions, while a second (three-cylinder, 240-horsepower), mounted on a raised deck at roof level, drove a generator to power the two electric swing motors. These two slow-revving engines, especially when lugging down under load, produced an unforgettable sound as they barked in and out of phase.

Page modernized its draglines beginning in 1954, when it introduced the first of its 700 series. The Model 752 was the most popular, and swung a standard 42-yard bucket. Many of these found homes in the midwestern coal mines and in Florida phosphate fields. The largest dragline built by Page was the Model 757 delivered in 1983 to the Obed Mine near Hinton, Alberta, now operated by Luscar Ltd. Weighing 4,500 tons, it carries a 75-cubic-yard bucket.

In 1988, Page Engineering Company was purchased by Harnischfeger Corporation (P&H) of Milwaukee, Wisconsin, providing Harnischfeger with a range of

The newest dragline to operate in North America at the time of writing is this P&H 9020, which started work at the start of 2000 at Luscar Ltd.'s Boundary Dam Mine, Estevan, Saskatchewan. The 98-yard machine represents a complete design makeover since Harnischfeger (P&H) purchased the Page Engineering Company and its dragline designs in 1988.

—Keith Haddock

Page modernized its draglines in 1954 with the launch of the 700 series. The most popular of these was the 42-yard 752, of which 27 were sold. Hopkins County Coal Company operates the one shown here at its Cimarron Mine in Kentucky.

—Keith Haddock

walking draglines to supplement its popular shovels. The first P&H walking dragline was sold to British Coal Opencast for work in northeast England. Commissioned in 1991, it was an updated version of the Model 757. Its walking system was changed to a cam-and-roller type similar to that used on Marion and Rapier machines. P&H has since further modified its draglines, and assembled four of its new 9000 series machines in Australia. The first Model 9020, which went to work in 1996 at the Bulga Coal operation in New South Wales, swings a 115-yard bucket on a boom 320 feet long. The newest 9020 started work in January 2000 at Luscar's Boundary Dam Mine, Estevan, Saskatchewan, Canada. It carries a 98-yard bucket on a 350-foot boom.

The world's largest draglines currently operating are the two Bucyrus-Erie 2570-WS machines. These are a "super" version of the 2570-W dragline, but they bear little resemblance to the earlier machines in either design or capacity. One has worked since 1993 at the Black Thunder Mine in Wyoming, swinging a 160-yard bucket on a 360-foot boom. The other, an 8,000-ton machine, started in 2000 at BHP Coal's Peak Downs Mine in Australia. This modern-day monster has a rated suspended load of 800,000 pounds, ranking it high on the dragline size scale. Only the two 3270-Ws and the 4250-W "*Big Muskie*" are larger than this modern-day monster. There are 34 main electric motors on board—8 hoist, 8 drag, 14 swing, and 4 to power the walking shoes. The Peak Downs machine is equipped with a 360-foot boom, and is designed to carry buckets in the 140- to 160-yard range.

Above: The largest dragline Page produced was this Model 757, *Athabasca Rose*, which swings a 75-yard bucket at Luscar Ltd.'s Obed Mine, near Hinton, Alberta. It was delivered in 1983.

Below: The world's largest operating draglines at the time of writing are two Bucyrus 2570-WS machines. The one shown here is *Ursa Major*, swinging a 160-yard bucket at the Black Thunder Mine in Wyoming. The other started work in 2000 at BHP Coal's Peak Downs Mine in Australia.

—Urs Peyer

CHAPTER SIX

THE GIANT DRAGLINES

By the beginning of the 1960s, walking draglines had become well established in the surface mining industry, and as machine size had kept nudging upward during the previous decade, mechanical and electrical problems were solved, and machines became more reliable. Ransomes & Rapier of England broke the world size record in 1951 with the launch of its W1400. By the end of the decade, both Bucyrus-Erie and Marion had equaled or surpassed its size with their 1250-W and 7800 machines, respectively. Then Rapier recaptured the title in 1961 with its record-breaking 40-yard W1800. Operators were realizing the advantages of large draglines, and they adjusted their mine plans to accommodate the longer ranges being offered by manufacturers. It appeared that the larger the dragline, the lower the cost-per-yard of moving the overburden. In addition, deeper coal reserves could be recovered. Thus the stage was set for something really big.

Big Muskie, the world's largest walking dragline, swings the biggest bucket ever to be suspended on a boom. At 220 cubic yards, the 4250-W Bucyrus-Erie machine is shown operating at the Muskingum Mine of the Central Ohio Coal Company soon after it first went to work in 1969.

—Eric C. Orlemann

109

The Marion 8800 was the world's largest dragline when it went to work in 1963. Carrying an 85-yard bucket, it was double the size of any dragline operating at the time. After the machine was proven, it was upgraded to carry a 100-yard bucket. Peabody Coal Company operated the 8800 at its Homestead Mine in Kentucky.

In 1961, Marion received an order for a dragline more than double the size of any dragline then in operation, and three times the size of its previous largest. Known as the Model 8800, it was ordered by Peabody Coal Company for its Homestead Mine in Kentucky. The new dragline was designed with a boom 275 feet long, swinging an 85-yard bucket. That's big enough to hold a small two-story house. The total width of the machine was 116 feet, as wide as a six-lane highway, and its rear end radius was 74 feet. The rotating frame was 15 feet deep and supported a gantry or 'A' frame that reached 115 feet above the ground. The circular base was 80 feet in diameter, providing a bearing area of 5,026 square feet for the working weight of 6,285 tons. Figure 6.1 shows the desk layout of the 8800.

These far-reaching dimensions allowed the machine to dig 12 stories below grade, swing its bucket a tenth of a mile, and dump it on top of a 14-story building. Each walking shoe was 70 feet long by 16 feet wide, and when in propel mode, the machine walked in 7-foot 8-inch steps.

Running such a behemoth digging machine required a lot of horsepower. A ground trailing cable fed AC electric power to two motor-generator sets, each of 10,000 horsepower. The generated DC current powered eight hoist motors, six drag motors, six swing motors, and four propel motors, totaling 12,250 horsepower.

Marion's director of engineering, J. F. Weis, described the 8800 in a paper presented to the American Mining Congress on May 7, 1962:

> *It is difficult, even for some of us who are familiar with large draglines, to visualize a machine of this size. It is estimated the machine will take 45,000*

engineering hours before design completion of our 85-yard dragline.

Many challenging problems have been solved during the course of the design. One of the major problems was the determination of an exceptionally rugged and simple means for walking. It became apparent early in our calculations that it would take an impractically large walking shaft if only one were used for a machine of this weight. Thus a two-shaft design was adopted. The mechanism for each side of the machine consists of two cranks, geared together, each of which is connected to a link. The opposite ends of these links are connected to common pivot on the walking shoe. As the cranks are rotated, the shoes move in a predetermined elliptical path. An associated problem connected with the walking machinery was the need for a very rugged supporting structure blended into the revolving frame. After considering many options, it was decided that the load

Another 100 yards of Kentucky dirt is on its way to the spoil pile, carried by the Marion 8800. Note the twin walking cranks, which share the load when the huge 6,285-ton dragline takes a walk. The record-beating Marion 8800 was the first of the super draglines born in the 1960s.

—Eric C. Orlemann

transmitted to the frame, while lifting the machine while walking, would be taken by two 15-foot deep 'I' beams running the full width of the revolving frame.

In pushing back the frontier in design of large draglines we have exceeded the limits of many standard sizes of equipment and components. To use the conventional number of drag and hoist ropes to transmit the enormous amount of power used on this machine requires wire rope beyond the size that could

normally be made by the manufacturer. In addition, the drums and sheaves for such a large rope size would have to be tremendous. To avoid this problem, we decided to use four hoist and four drag ropes. Each rope is 3 1/8 inches diameter, and has a breaking strength of 389 tons. These ropes still require quite large drums and sheaves.

A very important factor uppermost in our minds from the outset was the necessity of having plenty of reserve capacity in this large machine to minimize downtime and maintenance costs. (With millions of dollars at stake—far better to be overdesigned than underdesigned!) This important policy has been impressed upon the minds of all those now working on the project. Some examples of reserve strength built into the machine are:

- The cable loads are being kept comparatively low and the drums and sheaves are made large enough to obtain long cable life.
- Swing rails and rollers are of generous size to provide low point loads.
- Oversize bearings are used throughout the machine.
- All structural members are designed with comparatively low stresses for long service against fatigue.

As part of the policy to make the machine operate with a minimum of downtime and low maintenance cost, we are designing all the large gears for split construction so they may be removed in halves. Such split gears are not commonly used for excavating equipment but have been widely used in other applications.

With regard to the boom structure, we have come to the conclusion, after taking all factors into consideration, that the conventional boom design using 'T' section and pipe lacing is the most economical for this machine. While the 'T' section and pipe lacing are the largest we have ever used, they are entirely within present steel mill capabilities. Furthermore, to obtain

Before *Big Muskie* was built, its designers commissioned a painting to show their vision of the monster machine. In the foreground, the no-operator automatic Muskingum Electric Railroad hauls coal to the AEP Muskingum River Power Plant, some 15 miles distant.

—Bucyrus International Inc.

ample boom strength to take the high swing loads, we have spread the boom feet considerably beyond that dictated by a comparison with machines of present design. The swing power on this machine is so great that the loads resulting from swinging become the largest force in establishing the boom design, whereas in most cases, the hoist load is the predominating factor in determining boom design.

Instead of complicating the construction by supporting the boom with a mast, which we considered at first, we decided to use a very high gantry. Since the machine has a long revolving frame deck, the gantry can be made high and still obtain a good angle between the front and rear gantry legs.

As a fitting gesture to a machine that reaches the space-age heights in development, we are furnishing the operator's cab with such things as an electrically powered front window, a comfortably upholstered seat with 6-way electrically powered adjustment, beautiful interior design and an air conditioner.

The 8800 went to work in 1963 with great success. After the machine was fully proven, the mine management substituted the 85-yard bucket with one of 100 cubic yards capacity. That's a 17 percent increase, proving Marion's contention that it had overdesigned the machine for the benefit of its customer.

Bucyrus-Erie 4250-W *Big Muskie*

When a new walking dragline started to uncover coal at the Muskingum Mine near Cumberland, Ohio, in 1969, history was in the making. The dragline swung a bucket of 220 cubic yards, the

Aerial shot showing *Big Muskie* during the later stages of assembly. The stiff-leg derrick was the main lifter, with the smaller cranes loading parts onto trailers that brought the components within the radius of the derrick. Larger cranes were brought in to assemble major parts.
—Bucyrus International Inc.

largest ever suspended from a boom. No wonder this monster machine, known as *Big Muskie*, has become the most famous of the large stripping machines used in the surface mining industry. *Muskie's* size and capacity are so enormous it's difficult to comprehend. Its boom can suspend a bucket and material weighing 550 tons or 1,100,000 pounds. The machine's dumping radius is 302 feet, so in other words, it can dump material more than one football field away. Built by Bucyrus-Erie Company, the 4250-W, its true model designation, attracted much publicity when it first went to work. Press photographers snapped photos showing several vehicles in the bucket, a high school band playing in the bucket, and distant shots of people depicted as minute specks against the huge hulk of the machine. All this publicity was not surprising because *Big Muskie* was not just the largest dragline ever built, it exceeded the previous world's largest by over 50 percent. (The previous largest was the Marion 8900 at 145 cubic yards, which had gone to work a couple of years earlier.)

Even in an era of tall buildings and huge machines, the sheer size of *Big Muskie* astounded first-time visitors to the site. Perched approximately 120 feet above the coal seam, the dragline hoisted 325 tons of earth and rock with a single bite, swung 90 to 180 degrees and dumped it up to 600 feet away—all in a cycle time of about a minute! From one position, its working range covered an area equal to a 6-acre park.

Figure 6.2 shows the electrical motor arrangement and the house layout of *Big Muskie*. The house measured

The vertical hydraulic cylinders for propelling are shown on the side of the 4250-W *Big Muskie* during erection. Some of the many support cranes assigned to the job can be seen, including a Bucyrus-Erie 110-T truck crane with "Trigon" boom at right. The boom of the stiff-leg derrick can be seen in the upper left of the picture.

—Bucyrus International Inc.

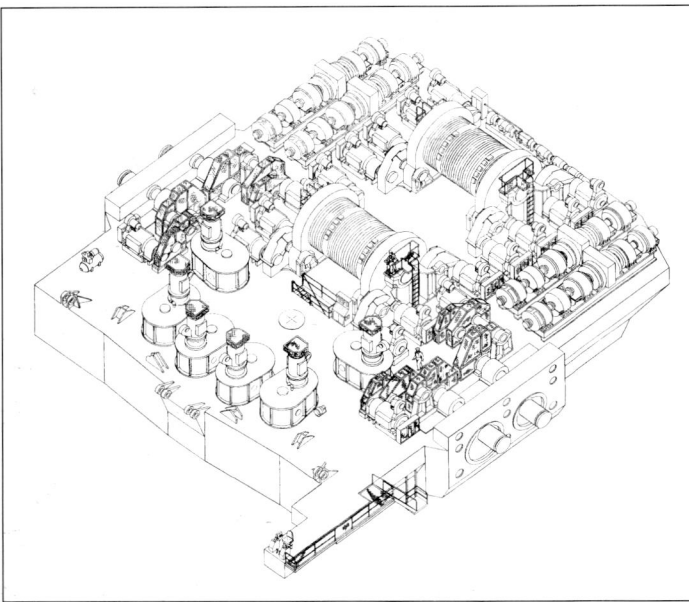

Figure 6-1

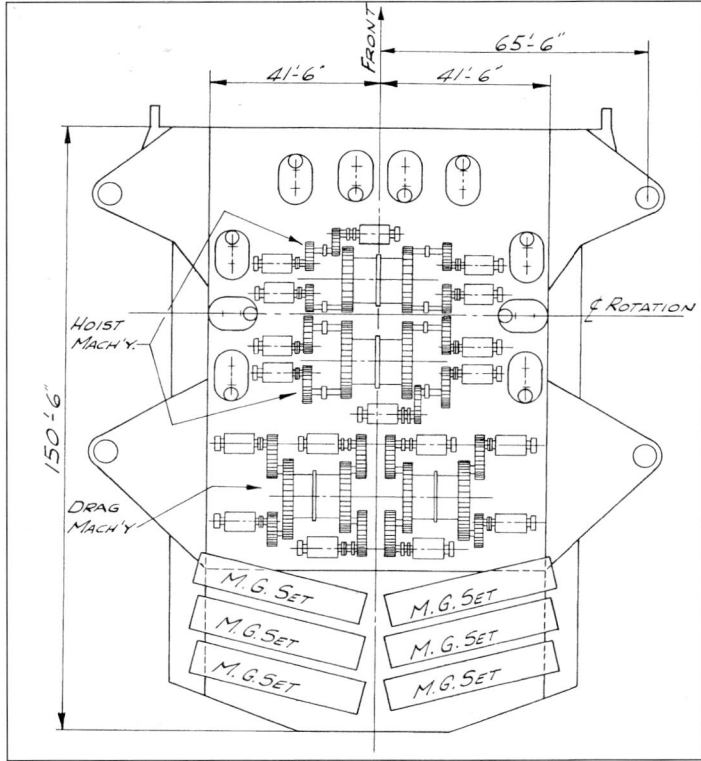

Figure 6-2

150 feet 6 inches long, 120 feet wide, and 40 feet high. Overall, the machine measured 487 feet in length, including the boom, and 151 feet in width over the walking shoes. The circular base or tub is 105 feet in diameter. Some quotations from a Bucyrus-Erie publication put these dimensions into perspective:

- Over three times longer than the Wright Brothers' first flight.
- Longer than 10 Greyhound Scenic Cruisers parked end-to-end.
- Almost 1-1/2 times the length of a football field.
- Wider than an eight-lane divided highway.
- Equivalent to the width of a football field.
- The 220-yard bucket can hold an average two-story house.

Design of the 4250-W started in 1965, four years before the machine dug its first bite. The Muskingum Mine owners, Central Ohio Coal Company, were in the market for a dragline of unheard-of proportions. They went to the only two manufacturers capable of building such a machine, Marion Power Shovel Company and Bucyrus-Erie Company. Both made proposals. Marion initially proposed two 110-yard machines, but that idea was not accepted. Central Ohio wanted a single humongous dragline, so Marion returned with its proposed Model 9600. In the end, the coal company chose Bucyrus-Erie's 4250-W, and two years later, the manufacturer was shipping parts to the erection site. It took 340 rail cars and 260 truck loads to ship the thousands of components. At the site, 300,000 man-hours of labor went into the assembly over a two-year span.

Big Muskie went into operation late in 1969. She ran with a crew of five, including the operator. He was assisted by an oiler, a welder, an electrician, and a laborer. In addition, two ground men operated bulldozers assigned to the machine and looked after the trailing power cable.

The 220-yard bucket was suspended by four of the biggest wire ropes ever manufactured at that time, measuring 5 inches in diameter. The four drag ropes and four hoist ropes had a total length of almost a mile. Hoisting was accomplished through two separate drums 11 feet in diameter, each winding two of the 5-inch ropes.

Ten 1,000-horsepower DC electric motors powered the hoist through pinions and bull gears. The drag motion had a similar system with its 11-foot drums powered by eight 1,000-horsepower motors. Another 10 625-horsepower DC motors powered the swing drive through separate gear reductions and pinions, which meshed with the 75-foot diameter swing gear. The resulting swing torque could accelerate the loaded bucket from 0 to 15 miles per hour in seven seconds.

Big Muskie was a veritable power house. Power was supplied by a 13,800-volt trailing cable, which had to be moved by special equipment. Self-propelled cable reelers wound the surplus cable as the dragline moved around the site. Muskie used as much electricity as 27,500 average homes. There were 293 AC electric motors in addition to the DC main motors already described. There were also 95 hydraulic pumps, 453 electric heaters, and over 90 1,000-watt mercury vapor lights, plus more than 500 smaller incandescent lights. All this lighting produced a daylight effect around the machine, as it worked 24 hours a day. From a distance, *Big Muskie* looked like a lighted Christmas tree in all its glory.

An important part of the electrical system was the safety network that monitored hundreds of conditions throughout the machine. There were some 120 automatic alarm systems that warned the crew whenever there was a fuse loss, an overheated bearing, a power failure, a lubrication malfunction, a hydraulic leak, or, most importantly, when a fire might develop.

To move about the mine site, *Big Muskie* used a unique walking system. Instead of the usual single shoe mounted on each side driven mechanically by cranks, four hydraulically powered shoes were used. Each of the four shoes measured 20 feet wide by 65 feet long. There were two on each side, pinned together to reduce distortion. They were raised and lowered by vertical hydraulic rams that lifted the base of the machine entirely off the ground, unlike the usual method in which the rear of the circular base trails on the ground. A further four horizontal hydraulic rams, two on each side, pushed the machine through its walking step when in the raised position. Each step was 14 feet, compared with 6 or 7 feet for other draglines. Walking speed was about 1/6 mile per hour, but like all draglines fed with a trailing power cable, the machine had to stop periodically to reposition the

A worker at Bucyrus-Erie's South Milwaukee, Wisconsin, plant proudly shows the comparative size of one of *Big Muskie's* bucket teeth. The hook at the tip is temporarily welded on to ease lifting.

—Bucyrus International Inc.

cable. The designers identified several advantages in the four-shoe walking system:

- The revolving frame can be supported at four points instead of two, thus reducing the bending moment in the frame.
- With the base entirely lifted off the ground, the sliding friction is eliminated.
- Sensing devices can vary the length of stroke of the lifting cylinders, so the amount of lift on each cylinder can be adjusted to suit ground irregularities.
- The length of step can be varied, as it is not governed by fixed crank dimensions.

A high school band is easily accommodated in *Big Muskie*'s 220-yard bucket in this publicity shot, taken soon after the dragline went to work in 1969.
—Bucyrus International Inc.

Big Muskie's boom was 310 feet long, and consisted of two separate square columns. Each column was built up from four 24-inch diameter tubes, laced together to form a single column unit, 18 by 20 feet in section, with the two columns joined together at the boom point. One Bucyrus-Erie engineer working on the boom compared it to designing a fishing rod for the Jolly Green Giant! Each time the machine went through an operating cycle, as is the case with all large draglines, the compressive and bending loads on the boom structural members were constantly reversing. The horizontal (side loads) on a dragline boom are greater than the vertical loads. That's why the boom feet on *Big Muskie* were placed 110 feet apart!

Maintenance or repairs on *Big Muskie*, where even the smallest component needed lifting gear, was a major operation. Electricians, mechanics, and welders were on duty round the clock to tackle any problem that might arise. One of the larger jobs included adding a large beam across the front of the revolving frame to prevent cracking in the frame. On another occasion, one of the two front fairleads broke off and fell to the ground. Twice during *Muskie*'s life, the tub was replaced. This was a maintenance job of gargantuan proportions, and took two years of planning and execution. To pull the tub in and out from under the machine, 32 major pieces of mobile equipment were hooked together, and all pulled as hard as they could!

Looking down on the massive structure of the 4250-W *Big Muskie* machinery house. The entire machine weighed about 14,500 tons.

—Bucyrus International Inc.

During operation, a Bucyrus-Erie 110-ton truck crane was assigned to *Big Muskie* to handle the many daily lifting tasks. The company kept a spare bucket on hand so Muskie could keep working when the original bucket needed maintenance or repairs. Moving one of these empty buckets on the ground took four loaded Caterpillar 657 motor scrapers. That's a total of 3,800 horsepower!

The operating weight of *Big Muskie* was never accurately determined. Bucyrus-Erie's published specification lists the weight as 27,000,000 pounds or 13,500 tons. But that was an estimated weight at the time of design. Since then the machine received significant modifications and structural additions, including a main support beam across the front of the revolving frame, and a heavier-designed tub. The final weight is believed to be just over 14,500 tons.

THE END OF *BIG MUSKIE*

Unfortunately the Muskingum Mine is located in a high-sulfur coal region. Increased concerns over acid rain, combined with the Clean Air Act Amendments of 1990, make this type of coal very expensive to burn because coal-fired plants must refurbish facilities to reduce emissions. The result is that coal contracts at the Muskingum Mine have diminished. *Big Muskie* was shut down in January 1991, and since

The Muskingum Mine

The mine where *Big Muskie* did its work was well suited to a machine of its size. Operated by the Central Ohio Coal Company, a division of American Electric Power, the Muskingum Mine is situated in hilly country, in rocky terrain, and has a relatively high mining ratio (approximately 27 cubic yards of earth and rock must be moved to produce 1 ton of coal). At its peak in the mid-1980s the mine was moving well over 100 million cubic yards of material, including salvage of surface soil and reclamation, and shipping 3.3 million tons of coal a year. To move such large quantities of material efficiently, Central Ohio Coal specified the largest mining machines.

With all this material to be moved, the mine amassed one of the largest earthmoving fleets ever to operate at one site. Before acquiring *Big Muskie*, the mine operated two 45-yard stripping shovels (Marion 5561), a 35-yard dragline (Marion 7800), an 18-yard shovel (Marion 5323), and two 12-yard draglines (Page 627 and Bucyrus-Erie 500-W). When *Muskie* went to work, all this equipment continued to operate for several years in different pits across the property. The 500-W was utilized to work ahead of the big dragline, cutting down local hills to make a working pad. With big strippers, it is important to keep them at full production in the main face and leave ancillary work to other equipment.

As the demand for coal grew, the mine added further major pieces of equipment and boosted its coal hauler and reclamation fleets. It brought in two new draglines, a 110-yard Marion 8750 dragline in 1980 and a Bucyrus-Erie 115-yard 2570-W dragline in 1989. Also in the mid-1980s, because overburden depth had increased, the mine purchased a fleet of four electric shovels and rear-dump trucks to reduce the hills ahead of the draglines. Taking as much as 60 feet of overburden in two benches, the "haulback" fleet as it was called, formed level working benches for the draglines. The name haulback was used because the material excavated from the pre-mined land was hauled back into the reclaimed area where the coal had been removed. The mine was proud of its reclamation program, which re-created productive land at the same rate untapped land was disturbed for mining. To salvage the soil ahead of mining and replace it on the recontoured land, the mine kept a fleet of 16 Caterpillar 657 motor scrapers at work.

The size of these rope sockets for the 5-inch diameter ropes gives some idea of the difficulties *Big Muskie* presented for maintenance crews.
—Keith Haddock

then, further reductions in equipment and major layoffs have followed. At the time of writing, the only dragline still operating at Muskingum is the Bucyrus-Erie 2570-W.

When *Big Muskie* was parked, she was in good operating condition. Over the years, all the operational problems, small and large, had been worked out, and many modifications had been performed by the mine's maintenance crews. During its life, *Big Muskie* moved more than 483 million cubic yards of material to uncover 18 million tons of coal. There was probably a lot of life still left in the machine, but with reclamation obligations and pressure to clean up the site, the owners decided to scrap her. Despite some efforts by a local preservation group to preserve *Big Muskie* as a mining interpretive center, there was not sufficient interest from the American public or governments to fund such an ambitious project. In 1999 a demolition contractor exploded the boom supports, and the boom fell to the ground. Several months later, *Big Muskie* was cut up and removed from the site. The amount of recycled steel was enough to produce over 9,000 cars!

As an epitaph to one of the world's greatest engineering feats, one of the 220-yard buckets has been saved. It now rests not far from where the once-proud *Big Muskie* made the earth move. Some years ago, a visiting Australian film crew was making a documentary on the American scene. Features included the Hoover Dam, the faces of Mount Rushmore, the Golden Gate Bridge, the World Trade Center, and *Big Muskie*. What a shame that the latter wasn't preserved in perpetuity, along with America's other top tourist attractions.

Big Muskie operating in 1990, nine months before it was finally shut down. The widely spaced A-frame, mast, and boom foot (110 feet wide) take the tremendous side loads as the dragline swings its 550-ton load. Compare the size of the Komatsu D475A bulldozer helping to smooth the working pad.
—Keith Haddock

APPENDICES

Appendix 1: A Brief Manufacturers' History

Companies that build, or have built, stripping shovels and/or walking draglines are detailed here.

Page Engineering Company
(Purchased by Harnischfeger Corporation [P&H] 1988)

The dragline as we know it today originated with Chicago drainage contractor Page & Schnable at the beginning of the twentieth century. John W. Page, a partner in the firm, is credited with building the first practical dragline machine. It was used on an excavation contract in 1904 on the Hennepin Canal in Illinois. Although some patents for machines resembling draglines existed as far back as 1880, there are no records of any actually being built before Page's. Page needed a machine to excavate below grade level, so he contrived a simple derrick mounted on a wooden frame with a wooden boom that could swing from side to side. The boom was controlled by winches. The steam-powered machine swung a 1-yard bucket, and the hoist and drag drums were worked by simple clutches.

Following the success of this first machine, Page built more draglines for use on its own contracts. One of these was the excavation for steel mill buildings in Gary, Indiana, in 1907, where some tough digging was encountered. Finding the standard hoisting machinery too light to operate his dragline successfully, Page ordered some specially built heavy-duty equipment from Monighan's Machine Works, a local Chicago firm that John Monighan established in 1884 as a general machine shop. This heavier machinery was so successful that other contractors placed orders for similar equipment. To meet the demand, Page and Monighan agreed informally to manufacture dragline excavators together. Monighan supplied the machinery and Page supplied the rest of the machine, including the bucket.

Eventually, building draglines proved more profitable than contracting, and Page Engineering Company was incorporated in 1912 to build draglines and dragline buckets. By this time, Page and Monighan were building draglines individually, although Monighan continued to supply hoisting machinery to Page until about 1916.

Page did not fit draglines with a walking device until about 1923. The walking system was very complicated, as described in chapter 5, and not a great success. But the vastly improved system unveiled in 1935 overshadowed any shortfalls in the earlier design. It was so successful that Page used it on every dragline until the company's demise in 1988.

During the Great Depression of the early 1930s, not many draglines were sold, but Page Engineering kept its employees busy rebuilding machines, taking in other work, and landscaping the grounds of the Page factory at McCook, Chicago. Good times would return.

In the 1950s, Page launched the redesigned 700 series draglines, which took the company to a prominent position in the industry, and provided other dragline makers with stiff competition in the size range up to 75 cubic yards. The company remained an independent family run business for its entire life. Founder, John W. Page, ran the company until his retirement in 1965 at the age of 96.

The company continued to build its popular 752 Model (42 yard class) and introduced the larger 757 (60 yard class) in 1977. The first of a new dragline model, the 852, was shipped in 1980. With many updated engineering features, it was intended to be the first of a new series of sophisticated draglines to replace the 700 series machines. Only one more 852 was built, however—and several more 700 series machines shipped—before the company closed its doors. Page delivered its last two draglines in 1986. They were Model 752s sent to a coal mine in Turkey.

In 1988 the company and its designs were sold to the Harnischfeger Corporation (P&H) of Milwaukee, Wisconsin, and the old Page factory was closed. For the first time, Harnischfeger could offer a line of walking draglines, and it continues to build them under the famous P&H trademark.

Monighan Machine Company
(Purchased by Bucyrus-Erie Company 1932, now Bucyrus International, Inc.)

Monighan's Machine Works was established in 1884 by John Monighan, a machinist with experience as chief of a railroad wrecking crew and shop foreman. Located in Chicago, his company started out as a general machine shop but soon expanded into manufacturing products such as coal stokers, mortar mixers, and hoisting engines. Within a few years, hoisting engines became the principal product of the firm.

Monighan became interested in manufacturing dragline excavators in 1907, when Page placed an order for hoisting machinery to be installed in a dragline. (See "Page" section above) Page was a local Chicago contractor who had built the world's first dragline in 1904. He was using one of his machines to excavate the foundations for a steelworks at Gary, Indiana, but found the standard hoisting machinery too light to operate his dragline successfully. So it was a natural move for Page to turn to Monighan, located in the same city, to supply a heavy-duty hoisting outfit. The heavier machinery gave excellent results and attracted the attention of other local contractors who wanted similar equipment.

After 1907, Page and Monighan agreed informally to build dragline excavators together. Monighan supplied the machinery; Page supplied his patented buckets and assembled the complete machine, including superstructure and boom. Both Monighan and Page accepted orders for the machines. In 1908, Monighan incorporated his firm as the Monighan Machine Company.

In 1913 a Monighan engineer named Oscar Martinson invented the first walking dragline, known as the Martinson Tractor (chapter 5), By that time, Page and Monighan had grown apart, and each was making draglines independently. Martinson's invention gave the Monighan machines an advantage. Martinson improved his walking system in 1925 by eliminating the suspension chains, and substituting a cam wheel running in an oval track in a frame pivoted to the shoes.

In 1927, John Monighan, now over 70 years old, declared his intention to retire from the business. A group of investment bankers subsequently purchased all of Monighan's interest, together with a portion of the holdings of the other owners, and established the Monighan Manufacturing Corporation as successor to the Monighan Machine Company. The former Monighan management group remained intact, and included Oscar Martinson, who was named president.

Although sales were strong leading up to the Depression years of the early 1930s, the uncertainty of potential weakening markets led Monighan management into discussions with Bucyrus-Erie Company on a possible takeover. A deal was struck in February 1931, whereby Bucyrus-Erie purchased the first of several blocks of shares in the Monighan company. Subsequent purchases of shares over the next three years gave Bucyrus-Erie a controlling interest in Monighan, and the company name was changed to Bucyrus-Monighan Company.

Under the new ownership, Monighan's Chicago plant and its management remained an independent unit and continued almost unchanged. One of the few changes that did occur was that steel castings were now purchased from Bucyrus-Erie's South Milwaukee foundries instead of from Chicago foundries. Sales of the very successful walking machines continued throughout the deep Depression of the 1930s, and Bucyrus-Monighan made a profit every year, a rarity for any company during that unforgettable period.

The formal merger of Bucyrus-Erie and Bucyrus-Monighan took place in 1946, and Bucyrus-Monighan's books were finally closed out by 1949. That same year Martinson retired from the company but continued working as a dragline consultant. The old Monighan Chicago plant continued to produce walking draglines as a separate unit under Bucyrus-Erie management. The plant was kept busy until 1957, when a recession in the mining industry led to its closure. Thereafter, walking dragline production was consolidated in Bucyrus-Erie's South Milwaukee plant.

Ransomes & Rapier Ltd.
(Purchased by Bucyrus-Erie Company 1988, now Bucyrus International, Inc.)

Ransomes & Rapier Ltd. was established in 1869 at Ipswich, England, by its parent company Ransome, Sims & Head, agricultural equipment manufacturers. The new company was founded by four former employees of the parent company, J. A. Ransome, R. J. Ransome, R. C. Rapier, and A. A. Bennett. They had an agreement to carry on the manufacture of railway and bridge equipment, which the parent company wanted to separate from its agricultural business. A new factory known as Waterside Works was built, and the company rapidly expanded to the height of railway development. It built an array of heavy machines, including locomotive cranes, large dockside cranes, water control gates, and concrete mixing plants.

The Ransomes & Rapier name soon became well known worldwide, with many industrial achievements to its credit. It made the first locomotive to run in China (1874), built the world's largest steam locomotive crane (120 tons at 20 feet radius), and constructed one of the world's first central concrete mixing plants in 1887, with elevators, storage bunkers, and rotary mixers. The company's water-control sluice gates were installed on many famous waterways, including the Nile in Egypt, Lloyd Barrage at Sukkur, India, the River Clyde in Scotland, and the River Thames and Manchester Ship Canal in England.

Ransomes & Rapier built its first excavator, a rail-mounted steam shovel shipped to Australia, in 1914. The machine was based on one of the company's standard locomotive cranes and featured a patented rope crowd of the automatic type—i.e., the same rope hoisted the bucket as well as working the crowd motion on the boom. After making these machines, the company saw little activity in the excavator field until 1924, when it made an agreement with the Marion Steam Shovel Company of Marion, Ohio, under which Rapier would build certain Marion-designed machines under license. Initially, some Marion machines built in the United States were shipped via New York to the Ipswich plant where they were assembled. The first all-British excavator was a Type 7 of 1 cubic yard capacity. These machines were sold as "Ransomes-Rapier-Marion" excavators. Over the next few years, Rapier progressively modified the Marion designs and added some entirely British-designed models in sizes not offered by Marion.

In 1934, Rapier shipped its first stripping shovel to an iron ore mine near Corby, England. It weighed 690 tons, and carried a 9-yard bucket on a 93-foot 6-inch boom. It was designated Model 5360, based on the Marion 360 shovel. The Rapier-Marion agreement terminated in 1936, but Rapier continued to build the stripping shovels with an increasing amount of British content in each one. Stripping shovel manufacture was discontinued during World War II, when the last one was shipped in 1944.

Rapier introduced its first walking dragline in 1939. The W170 was Rapier-designed throughout, and featured the patented Cameron & Heath walking device. The 4-yard machine was the forerunner of a full line of walking draglines produced by the company, including the world record-beating W1400 and W1800 machines of the 1950s.

In 1959 a major amalgamation took place between Ransomes & Rapier Ltd. and Newton-Chambers Ltd. of Sheffield, England, who had been making excavators branded "NCK" under license from Koehring Company of Milwaukee, Wisconsin, since 1947. A new company was formed, NCK-Rapier Ltd., which was owned 49 percent by Koehring and 51 percent by Newton-Chambers. In 1962 the joint management of the company decided to temporarily withdraw from the walking dragline business and concentrate on construction-sized excavators. The last of the old series machines, a W1800, started work in Italy in 1964. This machine was moved to Canada in 1973, and is still operating for Luscar at its Boundary Dam Mine in Saskatchewan.

From 1972, NCK-Rapier was owned by Central & Sherwood Ltd. which, in 1974, purchased the 49 percent share formerly owned by Koehring. With ties broken off with the American company, Ransomes & Rapier Ltd. assumed its old name and became an independent company once again. With the oil embargo crisis of the early 1970s, and a renewed worldwide appetite for walking draglines, the new Rapier management decided to reenter this market. The designs were revamped, and the first of the new series machines, a 44-yard W2000, was shipped in 1977 to C&K Coal Company in Pennsylvania.

Following a mining recession in the early 1980s, Bucyrus Europe Ltd., a subsidiary of Bucyrus International, Inc., purchased the manufacturing rights to Rapier's walking draglines in 1987. Based at the old Ruston-Bucyrus plant in Lincoln, England, Bucyrus still offers the W2000 Rapier-designed model in its dragline lineup. Because of its popularity in India, the company made a manufacturing agreement with Heavy Equipment Corporation (HEC), a government of India enterprise, allowing the draglines to be built locally in that country. At the time of writing, 22 of the W2000 series draglines have been ordered, the latest for shipment in 2000 and 2001 to coal mines in India.

Marion Power Shovel Company
(Purchased by Bucyrus International, Inc. 1997)

Henry M. Barnhart of Marion, Ohio, was a steam shovel operator in the early 1880s. He was continually annoyed by breakages and delays caused by design flaws in the machine he operated, an Oswego boom machine. He conceived a new type of shovel, to overcome most of the deficiencies found in the machines of the day. Needing financing and a place to build his machine, he turned to Edward Huber, founder of Huber Manufacturing Company of Marion, Ohio, builders of threshing machines and steam traction engines. Huber was impressed with Barnhart's shovel idea and immediately secured a joint patent with Barnhart in 1883. Known as "Barnhart's Steam Shovel and Wrecking Car," the first shovel was built in the Huber shops and sold to the Jackson & Mackinaw Railroad Company.

Barnhart operated and tested the shovel the first season, and he and Huber were so sure of its potential that, with another friend, George W. King, they founded the Marion Steam Shovel Company in August 1884. King was in business in his own right as a manufacturer of hay carriers. By the time the third shovel had been erected in the Huber shops, the young company was able to move into its own new factory in Marion, Ohio, the site that would remain the company's home for the rest of its existence.

The Marion company grew into one of the foremost manufacturers of excavating machines and, in competition with its archrival Bucyrus, made similar products, such as railroad shovels, dredges, cranes, walking draglines, and drills. Marion's vast excavator range extended from the smallest 1/2-yard shovel to the largest shovel ever put to work. This was the behemoth Marion 6360 at the Captain Mine, Illinois, with a 180-cubic-yard dipper capacity.

Marion achieved many notable "firsts" over the years. In 1911 the company put the first long-boom stripping shovel to work in North America, and followed with a succession of these giant machines, many breaking the world record for size. Marion entered the walking dragline business in 1939, and just three years later produced the largest dragline built up to that time. In April 1946, the company changed its name to the Marion Power Shovel Company to more closely reflect its products.

Marion expanded its construction-sized excavator business throughout the 1950s and early 1960s with some important acquisitions. In 1954, Marion purchased the Osgood Company and its subsidiaries: the General Excavator Company and Commercial Steel Casting Company, both of Marion, Ohio. It acquired the Quick Way Truck Shovel Company of Denver, Colorado, in 1961. Both these acquisitions enabled Marion to offer a broader range of cranes and excavators.

Marion made headlines when it built the famous Apollo moon rocket transporters for NASA in 1965. Based on stripping shovel undercarriage technology, the two diesel-electric transporters were designed to move fully assembled lunar spacecraft and rockets from the assembly building at Cape Canaveral to the launch pad, a distance of 3 miles. These huge vehicles weigh 3,000 tons without load, and are powered by six diesel generator sets producing some 7,600 horsepower. Still in use today, the transporters have taken part in most of the major space programs including the Space Shuttle.

Later in the 1960s, Marion gradually pulled away from the small machine market, preferring to concentrate on walking draglines, blast hole drills, and large two-crawler excavators above the 10-yard class. The large machine business was booming, and Marion stayed at the forefront by producing more record-breaking machines including the world's largest draglines in 1963, 1966, and 1967, and the world's largest shovel in 1965.

Marion's ownership changed hands a number of times over its long history. The owners are summarized below:

1884–1927	Privately held.
1927–1944	W. A. Harriman & Company, Inc.
1944–1954	Publicly held.
1954–1965	Merrit-Chapman & Scott Corporation. (Operated under subsidiary Universal Marion Corporation)
1965–1977	Pittsburgh Coke & Chemical Company.
1977–1992	Dresser Industries, Inc. (Operated as Marion Power Shovel Division)
1992–1996	Indresco, Inc. (Operated as Marion Division)
1996–1997	Global Industrial Technologies, Inc. (Name change from Indresco) (Operated as Marion Power Shovel Company)

In 1997, Marion Power Shovel Company was purchased by archrival Bucyrus International, Inc. (formerly Bucyrus-Erie Company). The merging of these two giants was a significant event in the earthmoving industry, and abruptly ended an intense competitive rivalry lasting 113 years. The plant at Marion, Ohio, was closed, but certain machines from the former Marion line have been updated and are available as Bucyrus machines.

Bucyrus-Erie Company
(Bucyrus International, Inc. from 1996)

The present Bucyrus International, Inc. and its direct forebears can boast a rich heritage of specialization in the crane and excavator industry. From the smallest yard crane to some of the largest machines ever to roam the earth, no other company can boast such a wide variety of excavating machines as those Bucyrus built over the past 120 years. Through its floating dredges, tractor equipment, hydraulic excavators, drills, cranes, walking draglines, wheel excavators, and other special machinery, the name Bucyrus has become synonymous with moving the earth.

A group of businessmen headed by Dan P. Eells, of Cleveland, Ohio, purchased the former Bucyrus Machine Company of Bucyrus, Ohio, and established the Bucyrus Foundry and Manufacturing Company in 1880. Its charter was " . . . to carry on the business of a foundry and of manufacturing machinery and railroad cars." In 1882 the company received an order for a steam shovel from the Ohio Central Railroad. The machine delivered was known as the "Thompson Iron Steam Shovel and Derrick," named after its designer John Thompson, manager of manufacturing at the company. The machine was a great success, and 59 were sold up to 1889. At this point, the company name was changed to the Bucyrus Steam Shovel and Dredge Company to reflect the company's increasing role as an excavator specialist.

In 1891, the company laid plans to construct a new factory at South Milwaukee, Wisconsin, where production commenced in 1893. Slow sales and financial difficulties led to the reorganization of the company, and a name change in 1896 to the Bucyrus Company.

Bucyrus shovels gained a solid reputation and sales increased. One of the company's most notable achievements was supplying 77 steam shovels to the Panama Canal project between 1904 and 1908.

Bucyrus acquired Heyworth-Newman draglines, designed by contractor James O. Heyworth, in 1910, giving the company a line of nonwalking machines mounted on either rails or skids and rollers. That same year, the company purchased a controlling interest in long-established competitor, Vulcan Steam Shovel Company, creating the Bucyrus-Vulcan Company. This company was short-lived, as the following year it was consolidated into a new Bucyrus Company, along with the Atlantic Equipment Company. The latter was established in 1902, and built well-designed, heavy-duty steam shovels used in mining.

An important date in Bucyrus Company history was 1927. That's the year Bucyrus-Erie Company was formed by the consolidation of Bucyrus Company and the Erie Steam Shovel Company of Erie, Pennsylvania. Erie had been building small steam shovels since 1914 as the Ball Engine Company, and had changed its name to the Erie Steam Shovel Company in 1922. The former Bucyrus Company had achieved great success with its large shovels and dredges, but its small machine market share was less impressive. The Erie merger enabled the company to offer a broad range of reliable machines from the smallest to the very largest.

In 1930, Ruston-Bucyrus Ltd. was established at Lincoln, England, with ownership split almost equally between Bucyrus-Erie Company and the British firm Ruston & Hornsby Ltd. which with its predecessors, had been building steam shovels since 1874. At the time of the merger, Ruston & Hornsby was the only company outside the United States offering a full line of excavators of all sizes, including the giant stripping shovels. The largest of these, at 10 cubic yards capacity, competed with those offered by Bucyrus. Ruston-Bucyrus consolidated the lines of both companies and, with free access to all Bucyrus-Erie designs, gradually replaced the Ruston models with American designs.

During the early 1930s, Bucyrus entered the walking dragline market by purchasing a controlling interest in the Monighan Manufacturing Corporation of Chicago, Illinois. Monighan had been making walking draglines with the unique Martinson propel system since 1913, and at the time of the takeover was one of only two companies in the world producing walking draglines. The other was Page Engineering Company, also of Chicago.

Bucyrus-Erie purchased the Armstrong Drill Company of Waterloo, Iowa, under an agreement signed in 1933. Armstrong could trace its manufacturing roots back to 1868, and offered a line of oil-well, water-well, and blast-hole mobile drills. These products were well suited to Bucyrus-Erie's existing customers, who often had to blast in rocky excavation sites. The drills were sold as Bucyrus-Armstrong products under a royalty agreement from 1933 to 1943, and thereafter as Bucyrus-Erie products when the agreement expired. Today, blast-hole drills are one of the company's three key products, along with shovels and walking draglines.

Bucyrus-Erie designed and produced tractor equipment (scrapers, bulldozer blades, and winches) from 1935 until 1954. Thousands of these successful units were shipped worldwide for use in World War II. They were usually fitted to crawler tractors made by International Harvester Company, with whom Bucyrus had formed an alliance.

Other important companies acquired by Bucyrus-Erie include the Milwaukee Hydraulics Corporation (1948) and the Hy-Dynamic Company (1971). The former, with its truck-mounted hydraulic crane, sowed the seed for what would become an extensive line of Bucyrus-Erie truck cranes, of up to 110 tons capacity. The latter company produced the Dynahoe, a tractor-mounted combination loading shovel and hydraulic backhoe, which also became a Bucyrus-Erie product, and remained in the line until 1985.

A recession in the construction industry in the early 1980s led Bucyrus-Erie to transform itself into a company serving only the surface mining industry. The size of the company was drastically reduced, and all manufacturing was consolidated at its South Milwaukee plant. All rights to its nonmining products, including truck cranes, offshore cranes, tractor-backhoes, and construction-size hydraulic and cable excavators were either sold to other companies or discontinued by 1985. The remaining products—electric shovels, walking draglines, and drills—became the company's main products, and were updated to serve the mining industry. As a special one-off design, Bucyrus built the largest-ever cross-pit wheel excavator and placed it in service at the Captain Mine in Illinois in 1986. This 25-story-high monster weighed 5,380 tons.

In an effort to counteract the cyclic nature of the mining equipment business, Bucyrus-Erie purchased Western Gear Corporation, a manufacturer of aerospace products, in 1981. To recognize this diversity, the company changed its name to Becor Western, Inc. in 1985, but the Bucyrus-Erie Company name was retained for its mining machinery division. Two years later, Western Gear Corporation was sold, and Bucyrus-Erie Company was purchased by B-E Holdings, Inc., a group of private investors.

In 1988, Bucyrus-Erie purchased the walking dragline business of the British firm Ransomes & Rapier Ltd. This business now operates under Bucyrus Europe Ltd. at Lincoln, England, and markets the entire line of Bucyrus draglines in countries not covered by the parent company.

Following some financial difficulties in the early 1990s, Bucyrus-Erie's private holders returned the company to public ownership, and in 1996 the company became Bucyrus International, Inc. The following year, Bucyrus acquired Marion Power Shovel Company in one of the greatest takeovers in excavator history. This ended an intense competitive rivalry between these two companies that had lasted 113 years and produced many of the biggest machines ever to dig in the earth. Updated versions of some former Marion machines are still available under the Bucyrus logo.

Harnischfeger Corporation (P&H)
(Purchased Page Engineering Company 1988)

Alonzo Pawling and Henry Harnischfeger established a manufacturing company in 1884 in Milwaukee, Wisconsin. From the very start, the now-famous P&H trademark was affixed to their small building. Pawling and Harnischfeger made a wide variety of products, from brewing equipment to overhead cranes. Pawling retired in 1911, selling his company interest to Henry Harnischfeger, who changed the corporate name to Harnischfeger Corporation a short while later.

The company's first venture into excavating machinery was in ladder-type trench diggers, commencing in 1910. The first P&H excavator, a fully revolving dragline, appeared in 1914. Then in 1918, Harnischfeger introduced one of the first gasoline-powered draglines, and these machines became one of Harnischfeger's key products in the 1920s. The company allotted much time and research toward new methods of manufacturing and innovative design. Unlike many of its competitors, steam power was not a priority with Harnischfeger. The company developed its first electric shovel in 1933, and became the industry leader in electric shovels, in terms of units sold, by the 1960s. P&H electric shovels have outsold all others in every decade since.

The industry recession in the early 1980s caused Harnischfeger to rely increasingly on its long-term partner and licensee, Kobe Steel of Japan, as a low-cost supplier of equipment to the world construction crane market. In 1986, Harnischfeger Industries, Inc. was formed as a holding company for the main product divisions, which today consist of Joy Mining Machinery (underground mining equipment) and P&H Mining Equipment. Although it continued to build construction cranes in the United States up to 1988, a diminishing market share caused Harnischfeger to sell its construction crane business to a group of managers in a leveraged buy-out that year. The new company, Century II, continued marketing its construction cranes using the P&H trademark. In 1994 this company was merged into Terex Lifting, along with other crane and excavator manufacturers in the Terex group.

In 1988, Harnischfeger purchased Page Engineering Company of Chicago, adding the walking dragline to the P&H Mining Equipment stable. Since then, the company has revised and improved the former designs, introduced new models, and found markets for its draglines in Jordan, England, Mexico, Australia, and Canada. The largest machine to date, the 9020, carries buckets up to 115 cubic yards, some 50 percent bigger than any machine produced by the former Page company.

In 1991, Harnischfeger acquired Gardner-Denver's rotary blast-hole drill product line. With the addition of these large mining-type drills to its already established lines of shovels and walking draglines, P&H Mining Equipment could now offer the same three main product lines as its chief competitor, Bucyrus International, Inc.

Ruston & Hornsby Ltd.
(Merged with Bucyrus-Erie Company 1930, now Bucyrus International, Inc.)

Ruston excavators originated in 1874 when Ruston, Proctor & Company of Lincoln, England, purchased the patents to James Dunbar's part-swing railroad-type steam shovel. A year later, the company sold its first shovel, known as the "Dunbar & Ruston Steam Navvy."

Joseph Ruston had established the Ruston, Proctor company soon after he negotiated a share in the firm of Burton & Proctor in 1857, when he was only 22 years old. He soon transformed the company from a builder of agricultural machines, including portable steam engines from 1851, into one of the most powerful industrial engineering companies in the United Kingdom. The company manufactured products ranging from steam engines and pumps to oilfield equipment. The main products were steam traction engines, steam rollers, and locomotives, but the shovel business also prospered. Beginning in 1887, over 70 Dunbar & Ruston steam shovels were employed on England's Manchester Ship Canal. Over the following six years, 54 million cubic yards of material were moved on this project, the first ever to use powered excavators in considerable numbers.

By 1889, Ruston, Proctor & Company employed 1,600 men, and further rapid expansion ensued. During World War I, the company produced all manner of war machines, such as guns and submarine and tank engines. It provided 30,000 sea mines under naval contracts, and built over 2,750 fighter aircraft for the war effort. After the war, Ruston acquired the Grantham, England, firm of R. Hornsby & Sons, and the new Ruston & Hornsby Ltd. was born in 1918. The Hornsby firm, builders of engines, already had many achievements to its credit. By 1896 it had produced the world's first oil-engine tractor and the world's first oil-engine locomotive. A Hornsby engine was installed on the Statue of Liberty in 1895 to provide power for illumination. Another powered Marconi's transmitter that sent the first wireless message across the Atlantic in 1901. Hornsby also built the world's first full-crawler tractor in 1905, and later sold its crawler patents to the Holt Tractor Company in the United States.

The new company expanded into new markets, including automobiles in 1920. But the well-built Ruston & Hornsby cars could not compete with mass-production methods at established car producers, and after 1,300 cars had been sold, production was discontinued in 1925.

Throughout all this diverse activity, Ruston & Hornsby's excavator business continued to flourish. Many new models of all types had been introduced since the turn of the century, so it was no surprise that the company designed and built some of the world's largest excavators in the 1920s. Beginning in 1923, large excavators were offered in three sizes, up to the 10-yard No. 300 (see Appendix 3). These machines were built either as stripping shovels or rail-mounted draglines, and several were shipped overseas.

In 1930 the shovel interests of Ruston & Hornsby Ltd. were merged with Bucyrus-Erie Company, and a new jointly owned company, Ruston-Bucyrus Ltd., was established. At the time of the merger, Ruston was offering 11 models of excavators, but over the next few years, all of these were phased out in favor of new machines jointly designed by both partners.

Appendix 2:
Extreme Mining Machines
100 Named Stripping Machines

Name	Model	Dragline or Shovel	Owner	Mine Name	State/Province/Country
Appalation Lady	1260-W	D	K & J Coal	Westover	PA
Bheem	7820	D	Singarini Collieries	Ramagundam	India
Ace of Spades	757	D	RJB Mining	Stobswood	U.K.
Achiever	680-W	D	R&F Coal	Polen	OH
Angeline	1250-B	D	Weirton	Hanover	PA
Anthracite King	1250-W	D	Pagnotti	Jeddo	PA
Athabasca Rose	757	D	Luscar	Obed	AB
Beulah Belle	500-W	D	Knife River	Beulah	ND
Bienfait Badger	1570-W	D	Luscar	Bienfait	SK
Big Bob	W2000	D	Miller	St. Aidans	U.K.
Big Digger	5960	S	Peabody	River Queen	KY
Big Dipper	1300-W	D	Colowyo Coal	Colowyo	CO
Big Dipper	1570-W	D	Sabine Mining	Hallsville	TX
Big Dipper	5761	S	Consol	Burning Star #2	IL
Big Divot	1370-W	D	Old Ben	No. 2	IN
Big Don	5700	S	Arch Coal	Captain	IL
Big Geordie	1550-W	D	Crouch Mining	Butterwell	U.K.
Big Hog	3850-B	S	Peabody	Sinclair	KY
Big Joe	8050	D	Adobe	Grove City	PA
Big John	1570-W	D	Hobet	No. 21	WV
Big John	1260-W	D	RJB Mining	Coalfield North	U.K.
Big Kate	2570-W	D	Old Ben Coal	Old Ben	IN
Big Lou	2570-W	D	Luscar	Boundary Dam	SK
Big Muskie	4250-W	D	Central Ohio Coal	Muskingum	OH
Big Paul	5760	S	Peabody	Hawthorne	IN
Big Red	1450-W	D	Old Ben	Blackfoot No.5	IN
Big Sandy	736	D	Baukol-Noonan	Center	ND
Bigfoot	1570-W	D	Luscar	Paintearth	AB
Brilliant Star	W2000	D	Brilliant Coal	Glen Allen	AL
Brutus	1850-B	S	P&M	No.19	KS
Brutus	8200	D	Luscar	Paintearth	AB
Brutus	7820	D	Southdown Cement	Wampum	PA
Captain	6360	S	Arch Coal	Captain	IL
Prairie Queen	1370-W	D	W.R. Grace	Four Corners	FL
Chevington Collier	1260-W	D	RJB Mining	Colliersdean	U.K.
Clementine	1350-W	D	W.R. Grace	Four Corners	FL
Coal Chief	5760	S	Peabody	Simco	OH
Crows Nest	8200	D	Morrison-Knudsen	Jim Bridger	WY
Dakota Star	762	D	Consol	Glen Harold	ND
Discovery	2570-W	D	Syncrude	Mildred Lake	AB
Doc	732	D	Green Coal	Panther	KY
Donald Duck's Digger	1300-W	D	Drummond	Beltona	AL
Elza	1570-W	D	Drummond	Flat Top	AL
Estevan Eagle	8750	D	Luscar	Boundary Dam	SK
Festus	480-W	D	Green Coal	Panther	KY
Freddie Flintstone	1260-W	D	Hallmark	Sipsey	AL
Gem of Egypt	1950-B	S	Consol	Egypt Valley	OH
Gentle Ben	1370-W	D	Old Ben	No.2	IN
Giant of Bedlington	1150-B	D	Parkinson	Ewart Hill	U.K.
Great Gus	2570-W	D	Luscar	Poplar River	SK
Green Hornet	5561	S	Consol	Mahoning Valley	OH
Groundhog	5561	S	Consol	Mahoning Valley	OH
King Coal	1370-W	D	Costain	Ravensworth	Australia
Lady of the Lake	8750	D	TransAlta	Highvale	AB
Lady of the Lake	1260-W	D	N.B. Coal	Chipman	NB
Little Beaver	480-W	D	Syncrude	Mildred Lake	AB
Little John	1260-W	D	RJB Mining	Coalfield North	U.K.
Little Sandy	380-W	D	Shand Mining	Apraw	IN
Long Tom	7800	D	Russell Coal	Cobb	AL
Master Miner	1250-W	D	IMC	Achan	FL
Missouri Quest	2570-W	D	Coteau Properties	Freedom	ND
Midway Princess	8200	D	P&M	Midway	MO
Big Kahuna	9020	D	Bulga Coal	Mt. Thorley	Australia
Miss Panther Valley	480-W	D	Greenwood Stripping	Nesquehoning	PA
Mountain Mover	8200	D	Addington	Martiki	KY
Mountaineer	5760	S	Consol	Egypt Valley	OH

Name	Model	Dragline or Shovel	Owner	Mine Name	State/Province/Country
Mr. Charlie	2355	D	Coal Systems	Oak Grove	AL
Mr. Dillon	1650-B	S	Green Coal	Panther	KY
Mr. Diplomat	950-B	S	Luscar	Diplomat	AB
Mr. Heman II	1570-W	D	Drummond	Mill Creek	AL
Mr. Jobe II	2570-W	D	Drummond	Cedrum	AL
Mr. Klimax	7800	D	Luscar	Boundary Dam	SK
Mr. Tom	1570-W	D	Drummond	Knobb	AL
Nessie	1300-W	D	Luscar	Sheerness	AB
New Horizon	8200	D	Hobet	No. 7	WV
Nicolaus Silver	5323	S	United Steel	Colsterworth	U.K.
Oddball	1150-B	D	RJB Mining	St. Aidans	U.K.
Pioneer	300	S	Hanna Coal	Mahoning Valley	OH
Prairie Queen	480-W	D	Knife River	Beulah	ND
Prairie Queen	1570-W	D	Luscar	Boundary Dam	SK
Prairie Rose	736	D	Luscar	Sheerness	AB
Priscilla	2570-WS	D	BHP	Peak Downs	Australia
Queen 'o Buttes	7620	D	Knife River	Gascoyne	ND
River Queen	1650-B	S	Peabody	River Queen	KY
Rosebud	1300-W	D	Peter Kiewit	Rosebud	MT
Sakakawea	2570-W	D	Coteau Properties	Freedom	ND
Sequoia	1570-W	D	Drummond	Kellerman	AL
Silver Spade	1950-B	S	Consol	Mahoning Valley	OH
Spirit of 76	757	D	Union Oil	Obed	AB
Spirit of Whitewood	8200	D	TransAlta	Whitewood	AB
Stripmaster	5761	S	Peabody	Lynville	IN
Sundew	W1400	D	British Steel	Harringworth	U.K.
Texas Star	1570-W	D	Sabine Mining	Hallsville	TX
Tiger	5561	S	Consol	Mahoning Valley	OH
Trail Blazer	550-B	S	Consol	Mahoning Valley	OH
Ursa Major	2570-WS	D	Arch Coal	Black Thunder	WY
Walking Stick	1300-W	D	Arch Coal	Black Thunder	WY
Wasp	5561	S	Consol	Mahoning Valley	OH
Whispring Thunder	9020	D	Peabody	Warkworth	Australia
Yei-Bi-Chai	1350-W	D	BHP	Navajo	NM

Appendix 3:
Extreme Mining Machines
Stripping Shovels

Make	Model	Dipper Range (CuYd)	Operating Weight (tons)	First Shipped	Last Shipped	No. Produced	Notes
Bucyrus	Class 5	1-1/2	90	1910	1911	3	Vulcan Steam Shovel design
	150-B	2-1/2	158	1912	1924	~17	
	175-B	3-1/2	220	1912	1925	~40	Plus approx.14 Class 175 draglines
	200-B	5–7	372	1927	1943	~13	Plus approx. 7 Class 200 draglines
	225-B	6	348	1915	1923	~90	Plus approx. 1 Class 225 dragline
	320-B	8	438	1923	1930	29	Plus 8 Class 320 draglines
Bucyrus-Erie	375-B	8	607	1931	1940	3	Plus 4 Class 375 draglines
	385-B	12	634	1929	1929	1	
	550-B	20–27	880	1936	1954	8	
	750-B I	12–16	922	1928	1930	10	
	750-B II	18–24	1,000	1930	1940	4	Counterbalanced hoist
	950-B	30	1,250	1935	1941	10	Counterbalanced hoist
	1050-B	33–45	1,523	1941	1960	12	Counterbalanced hoist
	1650-B	55–70	2,450	1956	1964	5	
	1850-B	90	5,220	1963	1963	1	
	1950-B (Spade)	105	7,200	1965	1965	1	Knee-action crowd
	1950-B (Gem)	130	6,850	1967	1967	1	Knee-action crowd
	3850-B (Hog)	115	6,950	1962	1962	1	
	3850-B (R/King)	140	9,350	1964	1964	1	
Marion	125	3	180	1925	1933	22	Plus 16 Type 125 draglines
	211	2	95	1913	1915	2	
	212	2	98	1916	1923	4	
	250	3-1/2	150	1911	1913	19	
	251	3-1/2	185	1913	1916	10	
	252	3-1/2	200	1916	1927	10	
	270	5	260	1912	1913	5	

Make	Model	Dipper Range (CuYd)	Operating Weight (tons)	First Shipped	Last Shipped	No. Produced	Notes
	271	5	293	1913	1917	9	
	300	6	350	1915	1923	74	
	350	8	585	1923	1941	35	Plus 12 Type 360 draglines
	5161	4	183	1937	1937	1	4-crawler mounting
	5320	8–12	725	1929	1934	7	Plus 1 Type 5320 dragline
	5321	10	780	1937	1937	1	
	5322	10	825	1937	1937	1	
	5323	11–20	1,033	1941	1961	9	
	5480	12–16	975	1928	1932	11	Plus 4 Type 5480 draglines
	5560 I	18	1,158	1932	1934	5	Counterbalanced hoist
	5560 II	32	1,550	1935	1937	4	
	5561	35–45	1,675	1940	1956	17	
	5600	15	1,550	1929	1929	1	
	5760	60–70	2,750	1955	1958	5	
	5761	60–75	3,788	1959	1970	16	
	5860	80	5,175	1965	1966	2	
	5900	105–110	7,250	1968	1971	2	
	5960	125	9,338	1969	1969	1	
	6360	180	15,000	1965	1965	1	
Rapier	5160	3-1/2	258	1936	1938	2	Four-crawler mounting
	5360	8	650	1933	1940	6	
	5361	11	706	1935	1935	1	
	5362	9	650	1937	1937	1	
	5363	9	650	1939	1939	1	
	5364	9	650	1940	1940	1	
	5365	9	700	1940	1940	1	
	5366	9	728	1941	1941	1	
	5367	9	728	1942	1942	1	
Ruston	135	4	198	1925	1937	4	Plus four No. 135 draglines
	250	6	246	1923	1923	1	Plus three No. 250 draglines
	300	10	392	1924	1930	4	Plus six No. 300 draglines
Menck	KRA	8-1/2	540	1927	1927	1	4-crawler mounting
NKMZ (CIS)	EVG 15.40	19	1,268	ca. 1960	*	*	
	EVG 35.65	45	2,920	1965	*	*	

Notes on the above table: Manufacturer's shipping date is shown. Operation date is later.
* Information not available.

Appendix 4:
Extreme Mining Machines
Walking Draglines

Make	Model	Bucket Range (CuYd)	First Shipped	Last Shipped	No. Produced	Make	Model	Bucket Range (CuYd)	First Shipped	Last Shipped	No. Produced
Monighan/ Bucyrus-Monighan	Martinson-Tractor	1	1913	1913	1	Bucyrus-Erie	180-W	4-1/2–6	1954	1965	16
	1-T	1	1913	1925	117		200-W	4-1/2–6	1945	1956	55
	1-1/2 -T	1-1/2–2	1922	1924	15		380-W	10–16	1978	1985	18
	2-T	2	1915	1922	26		450-W	9–12	1948	1954	11
	2-1/2 T	2-1/2	1922	1922	1		480-W	13–18	1955	1979	35
	3-T	3	1913	1925	65		500-W	12–14	1946	1959	9
	3-1/2 T	3-1/2	1922	1922	3		650-B	15–17	1946	1954	13
	4-T	4	1924	1925	6		680-W	34–40	1982	1988	5
	1-W	1	1926	1926	6		770-B	19–21	1954	1965	9
	1-1/2 -W	1-1/2	1926	1926	2		800-W	28	1966	1970	2
	2-W	2–2-1/2	1926	1938	25		950-B	12	1935	1935	1
	3-W	3–3-1/2	1925	1944	36		1150-B	20–25	1944	1950	17
	4-W	4	1926	1932	8		1250-B	25–38	1951	1958	8
	5-W	4–6	1934	1948	79		1250-W	35	1959	1963	6
	6-W	5–6	1926	1934	24		1260-W	30–40	1965	1990	27
	7-W	6–7	1942	1971	34		1300-W	33–47	1971	1985	12
	9-W	8–10	1938	1954	44		1350-W	45–60	1967	1977	10
	10-W	8–12	1934	1939	5		1360-W	50	1975	1976	2
	15-W	12–14	1940	1940	3		1370-W	58–65	1970	1984	38
	6150	6–8	1929	1932	9		1450-W	60	1963	1968	7
	6160	6–8	1932	1938	16		1500-W	70	1970	1971	2
	8160	8–10	1931	1931	3		1550-W	65	1968	1968	2

Make	Model	Bucket Range	First Shipped (CuYd)	Last Shipped	No. Produced
	1570-W	58–80	1973	1991	46
	2550-W	75	1964	1966	4
	2560-W	85–90	1969	1969	2
	2570-W	100–115	1971	1983	27
	2570-WS	140–160	1990	2000	2
	3270-W	176	1977	1977	2
	4250-W	220	1968	1968	1
Ruston-Bucyrus	3-W	2-1/2–3	1941	1955	15
	5-W	3–5	1939	1971	62
	380-W	10–12	1979	1985	6
	480-W	15–18	1965	1978	10
	1260-W	30–40	1976	1982	6
Rapier	W80	2–2-1/2	1939	1939	2
	W90	1-1/2–2-1/2	1943	1954	13
	W150	3–6	1944	1962	23
	W170	4	1939	1939	1
	W300	7	1957	1963	4
	W600	10–11	1953	1960	4
	W700	12–14	1981	1983	5
	W1350	33	1960	1961	2
	W1400	20–23	1949	1958	3
	W1800	25–43	1960	1962	4
	W2000	31–44	1977	2001	21
	W2100	40	1988	1988	1
Marion	7200	5–8	1939	1958	57
	7250	13	1983	1983	1
	7400	9–14	1940	1974	93
	7450	10–14	1979	1985	7
	7500	13–20	1970	1981	16
	7620	30	1974	1974	1
	7700	30	1969	1969	1
	7800	20–35	1942	1964	19
	7820	32–44	1970	1993	13
	7900	40–49	1962	1971	6
	7920	40–60	1970	1970	2
	8000	55–62	1970	1971	2
	8050	51–64	1972	1986	35
	8200	65–94	1973	1996	34
	8400	60–80	1969	1971	3
	8700	70–85	1963	1965	2
	8750	80–135	1971	1993	24
	8800	85–100	1963	1963	1
	8900	130–155	1966	1967	2
	8950	150	1972	1972	1
P&H	752	28	1990	1990	2
	757	65–75	1990	1991	2
	9020	75–115	1994	1999	5
Page	411W	1-1/2	1923	1931	14
	430	8	c.1930	c.1930	1
	615	3–4	1936	1940	4
	618	4–5	1941	1957	19
	620	5	1936	1938	8
	621	5–7	1939	1953	49
	625	8–10	1940	1953	26
	627	12	1946	1946	1
	630	8	1937	1937	1
	631	8	1946	1946	1
	634	10	1940	1940	1
	721	6–7	1954	1970	21
	723	8–10	1955	1956	2
	725	10–12	1955	1962	10
	726	11–13	1954	1955	4
	728	10–13	1956	1966	18
	732	16–20	1959	1979	13
	734	15	1959	1960	2
	735	20	1960	1960	1
	736	17–30	1969	1984	9
	738	22–26	1962	1963	2

Make	Model	Bucket Range	First Shipped (CuYd)	Last Shipped	No. Produced
	740	30–33	1965	1980	6
	747	40	1963	1963	1
	752	30–45	1965	1986	27
	757	52–75	1977	1983	7
	762	54	1969	1969	1
	852	45–48	1980	1981	2
UZTM (CIS)	Esh 10.75	13	1950	1951	3
	Esh 14.75	18	1951	1959	21
	Esh 14.65	18	1949	1950	2
	Esh 15.90	20	1959	1980	141
	Esh 20.90	26	1975	1997	62
	Esh 20.65	26	1953	1953	1
	Esh 20.75	26	1969	1971	3
	Esh 25.90	32	2000	2000	1
	Esh 25.100	32	1958	1971	3
	Esh 40.85	52	1977	1984	4
	Esh 65.100	85	1991	1991	1
	Esh 100.100	130	1976	1976	1
NKMZ (CIS)	Esh 1	4-1/2	1948	1952	*
	Esh 4.40	5	1951	1969	*
	Esh 5.45	6-1/2	1965	c.1977	*
	Esh 6.45	8	c.1977	*	*
	Esh 6.60	10	1957	*	*
	Esh 8.60	13	1959	*	*
	Esh 10.60	13	1965	*	*
	Esh 10.70	13	1964	*	*
Locomo (Finland)	Terasmies 3	1/2	1951	1951	1

Notes on the above table:
Each of the manufacturers listed in the table also made nonwalking draglines.
Manufacturer's shipping date is shown. Operation date is later.
* Information not available.

INDEX

Bucyrus Class 5, 19
Bucyrus Company, 18–20
Bucyrus Europe, 101
Bucyrus International, Inc., 8
Bucyrus-Erie Company, 8
Bucyrus-Erie stripping shovel models,
    80-B, 10, 11
    150-B, 20
    175-B, 20
    225-B, 20, 21
    320-B, 11, 23
    750-B, 23, 24, 26
    950-B, 24–26, 28
    1050-B, 27, 30
    1650-B River Queen, 17, 29
    1850-B Brutus, 30, 31
    1950-B Gem of Egypt, 32, 53–55
    1950-B Silver Spade, 32, 50–53
    3850-B, 29, 30
    3850-B Big Hog, 40–44
    3850-B, Lot 2, 43–46
    4850-B, 32
    4950-B, 32
Bucyrus-Erie walking dragline models,
    380-W, 99, 102
    480-W, 58
    680-W, 102, 103
    800-W, 59
    950-B, 83, 87
    1150-B, 88, 90, 91
    1250-B, 91
    1250-W, 88, 91, 109
    1450-W, 94
    1550-W, 91, 93, 94
    1570-W, 93, 94
    2550-W, 95
    2570-W, 59, 95, 99, 100
    2570-WS, 106, 107
    3270-W, 96, 97, 100
    4250-W Big Muskie, 96, 109, 112, 114–120
Bucyrus-Monighan Company, 87
Bucyrus-Monighan walking dragline models,
    5-W, 83, 89
    7-W, 89
    9-W, 89
    15-W, 84
    200-W, 89
    450-W, 89
    480-W, 89
    500-W, 89
    6150, 82
Grossmith, 8, 18, 19
Harnischfeger Corporation (P&H), 103
    9020, 103–106
John J. Wilson & Company, 18
Marion Power Shovel Company, 58

Marion Steam Shovel Company, 8, 19, 20, 26, 90
Marion stripping shovel models,
    250, 19, 20
    270, 20
    271, 20
    300, 21
    350, 11, 21–23
    5323, 13
    5480, 23, 25
    5560, 24, 25, 27
    5561, 27, 29
    5600, 24
    5760 Mountaineer, 28, 35–41
    5761, 7, 29
    5900, 32, 33
    5960 Big Dipper, 14
    6360 Captain, 30, 32, 35, 46–52
Marion walking dragline models,
    7200, 86, 90
    7400, 86, 90
    7450, 100–102
    7500, 60
    7620, 60
    7800, 87, 90, 109
    7820, 61
    8050, 99
    8200, 99
    8700, 95
    8750, 57, 98, 99,
    8800, 94, 95, 110–114
    8900, 93, 95, 96
    8950, 94, 99
    9600, 96
Martinson 3-W, 86
Martinson Tractor, 80, 81
Martinson, Oscar, 67, 80, 81
Menck & Hambrock, 8, 12
Monighan and Heworth-Newman, 80
Monighan Machine Company, 58, 81
Monighan walking dragline models,
    1-T, 80, 81, 84
    3-T, 81
    3-W, 82
    6-W, 86
    10-W, 87
    6150, 87
Novo-Kramatrsky Mashinostroitelny Zavod (NKMZ), 100
Otis, William S, 10
Page & Schnable, 79
Page Engineering Company, 58, 80, 81, 84, 102
Page, John W., 79
Page walking dragline models,
    411W, 85
    620, 85

    621, 102
    625, 102
    627, 62
    631, 79
    728, 63
    736, 64
    752, 103, 106
    757, 103, 107
    762, 65
Ransomes & Rapier Ltd., 8, 9, 26, 27, 58, 89, 93
Ransomes & Rapier stripping shovel models,
    5160, 10
    5360, 9, 26, 27
    5365, 11
Ransomes & Rapier walking dragline models,
    W-80, 89
    W-90, 85, 89
    W150, 85, 89
    W170, 84, 89
    W300, 66
    W600, 71
    W700, 102
    W1350, 72, 73
    W1400, 93, 109
    W1800, 90, 93, 109
    W2000, 98
Ruston & Hornsby Ltd., 8, 23, 24
    No. 135, 8
    No. 300, 23, 24
Steam railroad shovels, 10
Stripping shovels,
    demise of, 32, 33
    erecting, 15
    knee-action crowd, 27
    machine description, 9, 10
    operating crew, 13, 14
    operation of, 13–15
    shipping, 15
    shovel mountings, 10–12
    Ward Leonard control, 12
Taylor & Hubbard, 8
Terasmies 3, 77
Uralmash (UTZM), 101
    Esh 15-90, 75
    Esh 25.90, 76, 101
    Esh 100.100, 101
Vulcan Steam Shovel Company, 10, 18
    Class L, 18
    Model K, 18
Walking draglines,
    machine description, 61–67
    operating crew, 71, 73
    operation of, 70, 71, 74, 75
    walking systems, 67–69
    shipping, 75–77
Whitaker, 8
Wright & Wallace, 17